The Syntactic Recoverability of Null Arguments

The Syntactic Recoverability of Null Arguments

YVES ROBERGE

McGill-Queen's University Press
Kingston & Montreal • London • Buffalo

ISBN 0-7735-0732-9
Legal deposit first quarter 1990
Bibliothèque nationale du Québec

Printed in Canada on acid-free paper

This book has been published with the help of a grant from the Canadian Federation for the Humanities, using funds provided by the Social Sciences and Humanities Research Council of Canada.

Canadian Cataloguing in Publication Data

Roberge, Yves
The syntactic recoverability of null arguments
Includes bibliographical references.
ISBN 0-7735-0732-9
1. Romance languages – Clitics. 2. Romance languages – Topic and comment. 3. Romance languages – Grammar, Generative. I. Title
PC191.C4R63 1990 440 C89-090336-0

Contents

Preface

This book studies the syntactic properties of subject and object clitic pronouns in several Romance languages and dialects from the perspective of the Principles-and-Parameters framework in generative grammar, developed in recent work by Noam Chomsky and others. It analyzes the properties of languages with subject clitics, such as Standard French, and those with no subject clitics, such as Standard Italian, in a homogeneous way. It generalizes the same treatment to object clitics and to clitic-doubling constructions, achieving a unitary solution for many aspects of the syntax of Romance constructions which are at the center of recent debates.

A comparative study of various, rarely discussed French dialects (Québécois, Algerian), Italian, and Spanish dialects allows central claims to be made specifically concerning the null subject phenomenon but also more generally with respect to syntactic theory. The inclusion of a wider range of data leads to the conclusion that French should be analyzed as a null subject language like many of its sisters in the Romance family and to a better understanding of the nature of the parameters responsible for the cross-linguistic variation found in syntactic constructions insolving clitic pronouns.

This book is a revised version of my doctoral thesis written at the University of British Columbia. Revisions were also made based on the comments from three anonymous reviewers for the Aid to Scholarly Publications Programme and McGill-Queen's whom I would like to thank.

The work presented here has benefited from the contributions of many people. I would first like to express my gratitude to Michael Rochemont who has been of invaluable assistance for the two years I have worked on this book. I am also indebted to Guy Carden, David Ingram, Richard Kayne, Robert Levine, and Hervé Curat for their comments on an earlier version. For their comments, suggestions, and general support: Helen List, Marie-Thérèse Vinet, Anne Lobeck, the late Alfredo Hurtado, Guy Nephtali, Osvaldo Jaeggli, Marianne Adams, Jeff McGill, Denis Bouchard, and Anna-Maria di Sciullo. I would also like to thank Joan McGilvray and Jonathan Mead from McGill-Queen's University Press for their valued assistance.

My special thanks go to Diane Massam for everything...

During my period of tenure at the University of British Columbia I received financial assistance from that institution, the Social Science and Humanities Research Council of Canada, and the Fonds FCAC de la direction générale de l'enseignement supérieur du Québec.

The Syntactic Recoverability of Null Arguments

Introduction

Linguistic theory as it is understood here includes a highly structured theory of universal grammar (henceforth, UG) composed of fundamental principles and parameters. The task of the language-learner is, in a simplified sense, to set the value of the various parameters on the basis of available evidence.

Recent developments in linguistic theory have pointed out that it is necessary to avoid very language-particular syntactic rules in favour of more general and fundamental abstract principles if the goal of linguistic theory is to account for the formal properties of natural language and how these properties are acquired by the language-learner.

One example of this shift concerns the transformational component of the grammar. Earlier works in the generative framework multiplied the number of highly language-particular rules necessary to account for the various syntactic constructions found in a given language. These transformations had the general form given in (1), where the variables stand for some elementary constituents of a string (e.g. syntactic categories).

(1) $X - Y - Z \rightarrow Y - X - Z$

Aside from general conditions on the form of the transformations, the rules had a very specific character. In order to restrict the class of possible grammars, subsequent studies attempted to reduce the language-particular transformations to the more general rule called Move-Alpha, where Alpha is a syntactic category. It is thus permitted to move any category anywhere as long as the other principles of UG are not violated. It may also be the case that the movement of a category is forced by these principles. I return to a more detailed discussion of the various components of UG in 1.1. In this introduction I wish to concentrate on a particular condition on transformations assumed in earlier works and to trace its development. This condition is often referred to as the condition on the "recoverability of deletion."

Suppose that a transformation deletes a syntactic category in a given structural environment. Chomsky (1964) points out that a sentence from which an element is deleted is generally not ambiguous and that this follows from the "recoverable" status of the deleted element. He assumes that a deleted element is always recoverable because,

(2) A transformation can delete an element only if this element is the designated representative of a category, or if the structural condition that defines this transformation states that the deleted element is structurally identical to another element of the transformed string.

Chomsky (1964, 41)

In the sense just expressed, the condition on the recoverability of deletion is a condition on a theory of grammatical transformations. Such a condition could be formulated as in (3); see Chomsky (1965, 222).

(3) Only recoverable deletions are permitted in the grammar.

A deletion is recoverable under the conditions in (2).

Let us consider the following English sentences to illustrate how the condition on recoverability of deletion functions.

(4)a. Want some coffee?
b. Want to see a movie tonight?
c. Getting pretty excited, aren't yoû?
d. She finish her thesis yet?
e. Looks like an accident.
f. Turns out you can't do that in Texas.

These sentences all have one thing in common: some parts of the clause are missing, namely an auxiliary and a subject in (4a,b,c), an auxiliary in (4d), and a subject in (4e,f). Nevertheless Akmajian, Demers and Harnish (1984) for (4a-d) and Schmerling (1973) for (4e,f) note that sentences such as these are quite common in informal speech. The interesting fact concerning them is that the missing elements can be easily "recovered" without ambiguity. Thus the sentences in (4) are interpreted as equivalent to the one (5) and cannot mean what is expressed by the sentences in (6).

(5)a. Do you want some coffee?
b. Do you want to see a movie tonight?
c. You are getting pretty excited, aren't you?
d. Did she finish her thesis yet?
e. It looks like an accident.
f. It turns out you can't do that in Texas.

(6)a. Do they want some coffee?
b. Does he want to see a movie tonight?
c. We are getting pretty excited, aren't you?
d. Will she finish her thesis yet?
e. He looks like an accident.
f. She turns out that you can't do that in Texas.

In a theory that includes condition (3), we can assume that some type of recoverability is at work in interpreting the sentences in (4). Now, because these sentences are not ambiguous, it must further be assumed that some systematic operation is involved, as expressed in (2). In effect, it is not the case that any element can be freely missing even when unambiguous recoverability is possible. This is illustrated in (7) below from Akmajian, Demers and Harnish (1984).

(7)a. *Last night's party not went too well.
b. *Do want some coffee?
c. *Is failing his courses, isn't he?

The ungrammaticality of the sentences in (7) must be taken as an indication that the systematic operation alluded to cannot be invoked in (7). What happens there is not that the nature of the missing element cannot be recovered. It is more likely that the specifications relative to the systematic operation are not met.

In a theory including (2) and (3), the non-ambiguity of sentences such as the ones in (4a,b,d) would be attributed to the application of language-specific rules such as the ones in (8) from Akmajian, Demers and Harnish (1984, 304).

(8) Abbreviated Question Rules
Rule 1: To form an abbreviated question, the auxiliary verbs (*do*, *have*, and *be*) can be deleted, or else they can be contracted onto the subject *you*.

> Rule 2: In forming an abbreviated question, the second person subject *you* can be deleted as long as an auxiliary verb is contracted onto it.

Besides blocking the sentences in (7a,b), the rules in (8) have the further consequence of allowing the recoverability of the missing element(s). In other words, given (2), (3), and (8), (4a,b,d) can be interpreted only as in (5a,b,d) and not as in (6a,b,d).

The important point here is that recoverability is made possible, in the theory just outlined, through very specific rules applying only in a given language and in a given context.

Let us now consider the sentences in (9).

(9)a. Italian

Ho	trovato	il	libro.
have-1s	found	the	book

'(I) have found the book.'

b. Swahili

a	– li	– mi	– ona
SUBJ(3S)	PAST	OBJ(1S)	see

'(He) saw (me).'

The general comment concerning the sentences in (4) holds of the ones in (9), that is, some elements are (in a sense) missing but the content of these elements is recoverable without ambiguity. The sentences in (9) can therefore not be considered as equivalent in meaning to the sentences in (10a,b).

(10)a. He has found the book.
b. They saw you.

It seems obvious that something systematic is happening in (9) that permits missing elements, on one hand, and their recoverability, on the other hand.

To illustrate the position we are adopting with respect to (9), it is possible to establish a parallel between the sentences in (9) and (4c). From a certain view, (4c) is very similar to the sentences in (9): the syntactic subject is missing but it is, in a sense, still present in the sentence. In the tag question (4c), the subject *you* appears at the end of the sentence. In (9a), the subject is represented by the verbal affix representing subject-verb agreement.

The same can be said of an English sentence such as the one in (11).

(11) John wanted to leave before Mary did.

Part of the predicate associated with the argument *Mary* is "missing" from its usual syntactic position but the missing VP is still there, present in the same sentence.

The same holds of (9b) where the subject and the object are represented by the subject marker and the object marker on the verb.

We thus consider here that the possibility of having empty syntactic argument positions – such as subjects and objects in (9) where their content is recoverable in a very local sense – is a property of some natural languages. I will refer to this property as the "null argument" property.

In a theory including (2) and (3), a plausible account of the property illustrated in (9) would be to postulate the existence of a subject deletion rule as proposed in Chomsky and Lasnik (1977). In (9b) a further rule of object deletion would also be required. The form of such rule(s) renders the deleted subjects and objects recoverable.

But, as stated earlier, explicit language-particular syntactic rules have been avoided in recent work in the generative framework. This shift clearly has consequences for a theory of recoverability such as the one presented above especially with respect to (2). If the transformational component of the grammar is reduced to Move α, then, clearly, (3), which is a condition on the form of transformations, is without foundation and must be replaced by more fundamental and general principle(s) or condition(s).

As we will see in 1.2, the property in (9), namely the null argument property, is now often attributed to the availability of the pure pronominal empty category "pro" in languages like Italian and Swahili. Pro is simply the phonologically null counterpart of a phonologically realized pronoun. It is also widely proposed that the verbal agreement somehow permits the presence of pro.

Thus, it appears that no language-particular rule of deletion is involved in the generation of the sentences in (9). It follows that the recoverability of the missing elements is not attributable to (2) and (3) above. Something more general is involved.

Before examining more closely what is involved in the recoverability of the missing elements in (9) we can ask whether the type of recoverability illustrated in (4) is intrinsically different from the type found in (9). In other words, should (9) and (4) be analyzed in the same way? I believe that the answer is probably affirmative even though I have no direct evidence to put forward to support this claim. However, it is highly plausible that the shift from a language-particular subject (or object) deletion rule to the assumption that a universally available empty pronominal is involved should also provide an account for the missing elements in (4).

In the rest of this book the term "null argument" will be used to refer to missing subjects and objects such as the ones in (9) but not in (4).

The proposal that specific properties of natural languages be attributed to general fundamental principles must also apply in the case of recoverability of deletion. It should therefore be possible, in the light of the above discussion, to replace the statements in (2) and (3). I propose to include in the theory of UG the fundamental Recoverability Principle in (12).

(12) Recoverability Principle
An empty category must be licensed.

According to the Recoverability Principle, the presence of an empty category in a syntactic structure must, in a certain sense, be permitted. Note that the passage from (3) to (12) illustrates a general shift in the theory from conditions on rules to independent principles.

In the present period of linguistic research in the generative framework, it appears that the different empty categories can be licensed by different principles or theories of UG. A trace, which is the empty category left by application of Move α, is licensed by the Empty Category Principle of Chomsky (1981). This principle states that a trace must be "properly governed". The only condition on the presence of the empty pronominal anaphor PRO in a syntactic structure is that the position it occupies must be ungoverned. Pro, according to what we have seen, must be licensed by verbal agreement.

One of the goals of this book is to propose a formal characterization of how pro is licensed in various constructions. In other words, how does the Recoverability Principle apply to pro? It will be suggested that, as with the licensing of trace and PRO, government theory plays a central role in the licensing of pro.

Much of recent work in the Government and Binding framework is devoted to the study of the property illustrated in (9) and it is fair to say that the general mechanisms involved are rather well understood. In the first chapter of this book, I present the features of this framework that are relevant here (1.1). This presentation is closed in 1.2 by brief reviews of some of the theories that have been proposed to account for the missing elements in (9). The historical development follows the natural path leading from the belief that the recoverability involved in (9) was rule governed to the proposal that it is in fact a good example of syntactic recoverability, i.e. that it involves more general principles from Binding theory and Government theory. This section is followed in 1.3 by a discussion of some issues relating to empty categories.

Chomsky (1981) points out that the property allowing the missing elements in (9), namely verbal agreement, can also permit an NP associated with an object clitic to be empty and says (278): "a close relation is established between clitic-NP pairs and subject-verb agreement, and the condition of recoverability of deletion for pronouns is to a considerable extent accommodated within the Government and Binding theory." Chapter 2 is concerned with this statement (2.1, 2.2, 2.3) and some residual issues (2.4, 2.5). In particular, I examine the exact way in which clitic-argument pairs and verbal agreement-argument pairs are similar. Most of the discussion in chapter 2 is devoted to subject clitics, which have received less attention in the literature. I also examine the symmetry existing between subject clitics and object clitics with respect to the syntactic structure and the relationship holding between these two types of clitics and empty arguments.

In the third chapter, I focus on the clitic-overt NP pair generally referred to as "clitic doubling." I explore the consequences of the theory developed in the second chapter for problems related to clitic doubling. The main topics addressed concern subject/object (a)symmetries with respect to doubling (3.1, 3.2) and the parameterization of the extraction possibilities out of subject doubling and object doubling constructions (3.3). This chapter ends with an examination of some apparently problematic constructions, namely, Complex Inversion (3.4) and French reflexive *se* (3.5).

The first part of the last chapter provides some proposals regarding the syntactic operation involved in the cliticization of subject clitics onto the verb (4.1). In the second part, I explore the possibility of analyzing clitics and agreement markers as a single syntactic notion. The conclusion reached there is that they are different entities even though they have the same role in the grammar: that is, clitics and agreement markers are syntactic devices necessary for the syntactic recoverability of null arguments.

CHAPTER ONE

Syntactic Theory and Null Arguments

1.0. INTRODUCTION

I present here the theoretical framework on which the different analyses are based. In the first section emphasis is put on the system of rules and principles constituting the core of the Government and Binding framework as developed in Chomsky (1981). Some of the modifications to this system presented in Chomsky (1982), in particular the one concerning the introduction into the framework of the empty category corresponding to an overt pronominal, and in Chomsky (1986b) are assumed as well. The aim of the second section is to introduce the reader to different analyses of the so-called null subject property. We will see that these analyses have focused on the government of the subject position by INFL and on the type of empty category occupying the subject position.

In the first part of this chapter, I present a discussion on the relationship between learnability and cross-linguistic variation as it is conceived of in the framework adopted in this book.

1.1. VARIATION AND LEARNABILITY

The way in which a specific theory is used is dependent upon how the language faculty can be described. Two major tasks must be handled: language development or growth in the individual (learnability) and grammatical variation among languages and dialects. The two are intimately related. It is generally the case, as we will see, that the explanation provided for one has consequences for the other. For example, if one wishes to account for a certain difference between two dialects of the same language and does so by providing a parameter that, depending on how it is fixed in the grammar, can explicate the difference, then one must also account for how the language learner will be able to fix the parameter on the basis of positive evidence.[1] That is, the formulation of the parameter must take into account the accessibility of particular constructions that must be used for its setting. Following Lightfoot (1982), we call this type of evidence a "trigger experience." In turn this constraint on the availability of a trigger experience has the effect of limiting the range of possible analyses. In other words, a proposed parameter that is unlearnable, i.e. for which no plausible trigger experience is available, should be rejected from the grammar since there is no way in which it can be fixed in one way or the other. Many examples of this relationship between variation and language growth will be presented in the following chapters.

Now, it is not obvious that this relationship can be handled straightforwardly by any theory. How then is the theory assumed here accomplishing this? According to Noam Chomsky, Jerry A. Fodor, and others, the central problem to be accounted for is how we come to know so much about language with so little evidence; cf. Chomsky (1986a). In other words, how can one reach the steady state that represents the adult grammar while being exposed to deficient stimuli. Katz (1966), Lightfoot (1982), the introduction to Hornstein and Lightfoot (1981), Chomsky (1975;1980b), and Roeper and Williams (1987), among others, argue that a child attains a knowledge of the rich and complex structures and rules of its language that goes far beyond what can be deduced from its linguistic environment. They thus assume that some kind of innate knowledge is available. This knowledge, represented by what is called Universal Grammar (UG), must have some specific properties in order to be included in a plausible research program. It has to be general and abstract enough to allow for all and only the possible particular grammars of the (possible) natural languages. UG must further include some notion, such as parameter, that has two (or more) possible settings to be fixed on the basis of positive evidence by the language learner. Alternatively, a parameter can already be set in a given way in UG. This particular setting constitutes the "unmarked" case. Positive evidence is needed to modify this setting to the "marked" case. Concrete examples of this type are presented in the following chapters.

A human being starts off at an initial state that constitutes the set of genetically encoded principles of UG and goes from one state to another on the basis of the different types of parameter settings until a steady state is reached – the adult grammar; cf. Chomsky (1980c). The theory of UG, as presented here, makes explicit the assumption that the genetic code, as opposed to the linguistic environment alone, is the main tool in the evolution from one state to the next. In fact,

Fodor (1980) argues that if we accept the idea that learning is the passage from some set of beliefs about the surrounding world to another set of beliefs by inductive inference, then it is logically impossible to acquire a richer system on the basis of a weaker one unless one assumes that this evolution is "mediated by the availability of any concept that can eventually turn up in the organism's beliefs" (Fodor 1980, 152).

In order to make clear the relationship between this conception of the growth of language in the individual and the variation among the languages of the world, it will be useful to illustrate such a conception with an example. The transformational component of a grammar, which has the universal form Move α in UG, is available to any language learner in the initial state. In other words, one does not have to "learn" that categories can be moved but only which category or categories can move.[2] A trigger experience in this case could be composed of two sentences forming a minimal pair, say, a sentence with a *wh*-phrase in situ and a sentence with a *wh*-phrase in clause initial position. From these two sentences it can be deduced that *wh*-elements undergo movement. Move α becomes more specific, namely Move *wh*. As to why, where, when, and how Move *wh* applies, other general principles of UG will provide the necessary answers. Suppose that the type of trigger experience above does not arise, then the possibility of moving *wh*-elements is not even considered. It is now clear that the discussion above distinguishes between two types of languages: those with *wh*-movement in syntax (English, French, etc.) and those without (Japanese, Chinese, etc.). That is the sense in which the principles and parameters of UG can reflect cross-linguistic variation; cf. Lightfoot (1982) for other examples.

Many instances of this type of relationship are presented in the following chapters, especially with regard to how languages differ with respect to the recoverability of syntactic arguments. But first, let us see in more detail the main modules of UG assumed in the rest of the book.

1.1.1. The Government and Binding Framework

1.1.1.1. System of Rules. The main goal of Chomsky (1981) is to provide an analysis of the distribution of NPs and show how a modular approach to grammar can achieve this task. This particular character of the grammar is captured by the rule system in (1).

(1) i. lexicon
    ii. syntax

a. categorial component
b. transformational component
iii. phonological form component
iv. logical form component

This system is organized in UG as shown in (2).

(2)

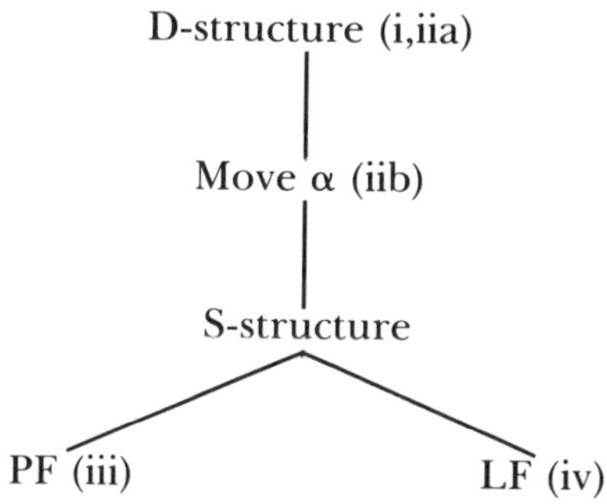

It follows from (2) that S-structure representations feed two independent components: Logical Form (LF) and Phonological Form (PF). There is no transparency at this level of the grammar, i.e. an operation taking place, say, in PF has no influence on the operations taking place in LF. I will have more to say on the properties of these two components below and in chapters 3 and 4.

The base-component is composed of the lexicon which includes specifications concerning the morphological structures of the lexical items; these specifications are expressed in terms of the X′-system; cf. Chomsky (1972;1986b), Jackendoff (1977), Stowell (1981) and others for proposals to dispense with phrase-structure rules. D-structures are thus produced by inserting lexical items in the syntactic representations produced by the X′-system. Following Chomsky (1986b), I assume the following major expansions:

(3) COMP″ → Spec C COMP′
COMP′ → COMP INFL″
INFL″ → Spec I INFL′
INFL′ → INFL VP

The Spec COMP position is the landing site for moved *wh*-elements whereas the head COMP of COMP″ (or CP) contains the complementizers, for example *that* and *for* in English. The subject of INFL″ (or IP) occupies the Spec INFL position; it is often referred to as the external argument of the verb.[3] I use NP dominated by I″ to represent this subject. The INFL node contains the AGR(eement) and TNS (tense) nodes; see 4.1 for more on the internal structure of INFL. The tree representation of (3) is given in (4).

(4)

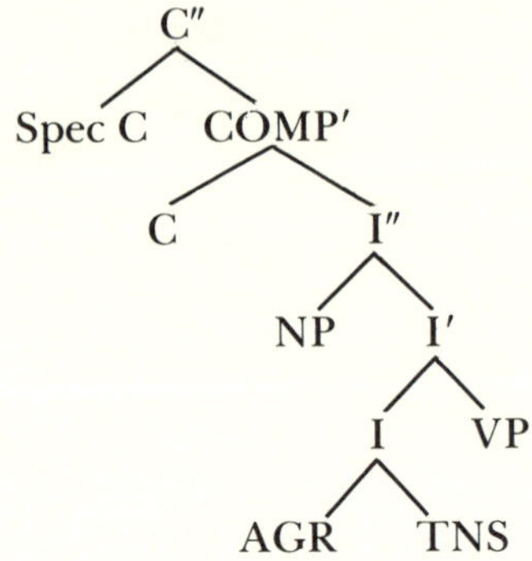

The highly rule specific transformational component of the earlier versions of the model is reduced in Chomsky and Lasnik (1977) to the unique rule Move α. The result of the application(s) of this rule to D-structure is S-structure. A moved element always leaves a coindexed trace of itself; cf. 1.3. Move α also applies in PF and LF. PF movements only affect the superficial order of the constituents and have no effects on the semantic interpretation of the syntactic structure. In LF, Move α is used, for example, to raise quantifiers and *wh*-elements. Contrary to the case in PF, LF movements affect only the interpretation of the structure and not the superficial order of the constituents; cf. May (1977) and Chomsky (1981).

1.1.1.2 System of Principles. The system of principles constitutes the internal organization of a grammar. The GB model is composed of a number of these principles which interact with the different levels and components illustrated in (2) above. These principles are the following:

(5)a. Bounding Theory
b. Government Theory
c. Theta Theory
d. Binding Theory
e. Case Theory
f. Control Theory

Bounding theory accounts for locality conditions. Subjacency is one such condition, which constrains the applications of Move α; it is not a condition on representations; cf. Chomsky (1982) for a discussion. It is defined in (6).

(6) Subjacency
In the structure:
X. . .$[_a$. . .$[_b$. . .Y. . .]. . .]. . .X
no rule can relate X and Y if a and b
are bounding nodes.

(Chomsky 1973)

Sportiche (1981), following Rizzi's (1982) work on Subjacency in Italian, demonstrates that the bounding nodes in Standard French are C″, NP and PP. His arguments are based on contrasts between sentences involving movement. Some examples are given in (7)-(9) with *wh*-movement.

(7)a. Tu connais l'homme qui a vu *x*.
'You know the man who saw x.'
b. *Qui$_i$ connais-tu $[_{NP}$ l'homme $[_{C''}$ qui a vu t_i]]?
'Who do you know the man who saw?'

(8)a. Il croit le fait que tu as vu *x*.
'He believes the fact that you saw x.'
b. *Qui$_i$ croit-il $[_{NP}$ le fait $[_{C''}$ que tu as vu t_i]]?
'Who does he believe the fact that you saw?'

(9)a. *Voila quelqu'un à qui$_i$ $[_{I''}$ je sais $[_{C''}$ lequel$_j$ $[_{I''}$ je crois $[_{C''}t_i$ que $[_{I''}$ j'offrirais t_j t_i]]]]]
'Here is someone to whom I know which one I think I will offer.'
b. Voila quelqu'un à qui$_i$ $[_{I''}$ je crois $[_{C''}t_i$ que $[_{I''}$ je sais $[_{C''}$ lequel$_j$ $[_{I''}$ j'offrirais t_j t_i]]]]]
'Here is someone to whom I think that I know which one I will offer.'

In (7b) and (8b) the trace cannot be related to its antecedent since they are separated from each other by two bounding nodes, namely C″ and NP. The contrast in (9) shows that C″ and not I″, as in English, is a bounding node. This is so because *à qui* in the ungrammatical sentence (9a) crosses two C″ nodes and two I″ nodes. In the grammatical sentence in (9b), *à qui* crosses two I″ nodes but only one C″ node. If I″ were a bounding node in French then we would expect (9b) to be ungrammatical.

The notion Government is central in the GB framework and it plays a major role in the discussions of the following chapters. Since the first attempts at defining this concept (Rouveret and Vergnaud (1980) and Chomsky (1980a)), many modifications have been proposed. I adopt here the definition given in Aoun and Sportiche (1981) as modified in Chomsky (1981, 165) without reviewing the different alternative proposals.

(10)a. Government
α governs δ in a structure $[_{\beta} \ldots \delta \ldots \alpha \ldots \delta \ldots]$ where,
i. $\alpha = X^{o}$
ii. φ is a maximal projection, if φ dominates δ then φ dominates α
iii. α c-commands δ

b. c-command
α c-commands β iff the first maximal projection that dominates α dominates β and α does not dominate β.

In general terms, government is defined as the relationship holding between the head of a projection and the projections that "depend" on it. We can simply say that a head governs its complements. It can be deduced from (10) that maximal projections are absolute barriers for government.[4] In (11), X cannot govern the nodes dominated by Y″ if Y″ is a maximal projection.

(11)

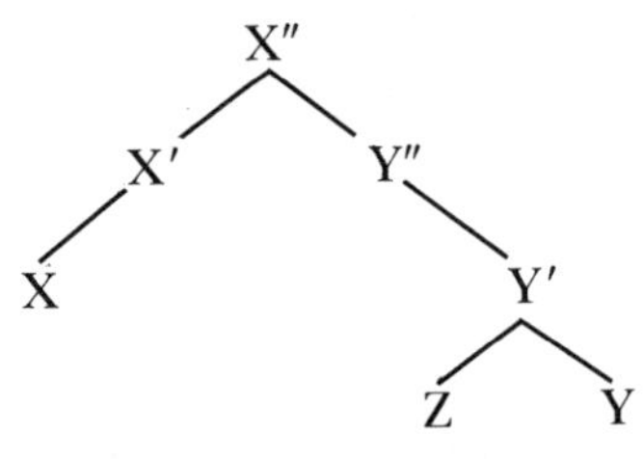

Government theory also includes the Empty Category Principle (henceforth, ECP) of Chomsky (1981). According to the ECP, a trace must be properly governed; cf. also (Kayne (1981a; 1981b)). Theta-theory aims at describing the thematic relations holding between arguments and predicates. Informally stated the Θ-Criterion requires that an argument in a Θ-position be assigned a Θ-role determined by its grammatical function. The Θ-role of an argument is determined by the Θ-position it occupies at D-structure. In (12) the trace of *John* but not the NP *John* occupies a Θ-position assigned a Θ-role.

(12) John_i seemed [t_i to like the cheesecake].

The Θ-Criterion is defined as in (13) in Chomsky (1981, 335).

(13) Theta Criterion
Given the structure S, there is a set K of chains, $K = \{C_i\}$, where $C_i = (\alpha_1{}^i, \ldots, \alpha_n{}^i)$, such that:
(i) if α is an argument of S, then there is a $C_i \in K$ such that $\alpha = \alpha_j{}^i$ and a Θ-role is assigned to C_i by exactly one position P.
(ii) if P is a position of S marked with the Θ-role R, then there is a $C_i \in K$ to which P assigns R, and exactly one $\alpha_j{}^i$ in C_i is an argument.

Binding theory makes crucial reference to government in the sense that the domain where a binding relation applies is defined under government. The domain in which a category is governed is another category which contains both the governor and the governee, namely the governing category. The three Binding Conditions are given in (14).

(14) Binding Conditions[5]
A. A [+anaphor] element must be A-bound in its governing category.
B. A [+pronominal] element must be A-free in its governing category.
C. An R-expression (variables, names) must be free.

Reference to A-binding indicates that the Binding Conditions cannot be reduced to a mere condition on coindexation that would say, for example, that an anaphoric element is free if it is not coindexed, and bound otherwise. Rather, A-binding refers to binding by an element in argument position. It follows that an element is "bound" if it is coindexed with an A-position that c-commands it.

Case theory was introduced and developed in Rouveret and Vergnaud (1980) and Chomsky (1980a). Case theory can be interpreted as a realization of the relationships among elements in a syntactic structure. It contains two important parts: Case-assignment contexts and a well-formedness condition on NPs, the Case Filter. It is generally assumed that Case is assigned to an NP under government by [+TNS], P, and V. Although very simple in its formulation, the Case Filter plays an important role in the grammar.

(15) Case Filter
*NP [−Case] if NP has a phonological matrix.

The filter applies at either S-structure or PF but after syntactic movement and essentially expresses the notion that a syntactic structure containing a non Case-marked, phonologically realized NP is ungrammatical.

Control theory provides the means to interpret the reference of the [+pronominal, +anaphor] empty element PRO. The mechanisms required for the interpretation of PRO, as opposed to its syntactic licensing, are not directly relevant for the discussions included in this book.

1.1.1.3. Interactions. It is now possible to assume that the interactions between the various elements of a sentence can be described by the principles listed in (5) and the model in (2). This approach to grammar enables us to reduce the system of rules to a minimum in number and complexity since the principles act as regulators which eliminate the possibility of over-generation.

Consider for example the transformational component. It is conceived of as the option of moving some element(s) anywhere in the sentence. In English, the element can be an NP marked with the feature [+*wh*] or not. Examples of this are given in (16).

(16)a. ϕ seems [Peter to have left].
b. Peter seems to have left.
c. Mary made $NP_{[+wh]}$.
d. What did Mary make?

In (16b) the lexical NP *Peter* (the subject of the lower clause in (16a)) is moved to the subject position of the higher clause. In (16d) an NP marked [+Wh] is moved from the direct object position in (16c) to a pre-sentential position.

We can ask the following questions: A) What forces the raising of *Peter* in (16a,b)? and B) Why is *Peter* moved to a specific position? or, why can't it move somewhere else in the sentence? The concept of grammar adopted here provides the following answer to A by appealing to certain principles. Case theory includes, as we have seen, the Case Filter as formulated in (15). In essence, the filter blocks any occurrence of a lexical NP that is not Case-marked. If we assume that in an infinitival clause the subject position does not get nominative Case, then it would be impossible for *Peter* to remain as the subject of *to have left*. That is the reason why *Peter* is forced to move somewhere else, i.e. in order to be Case-marked. In answer to B, the only possible landing site is the subject position of the higher clause. That is because a verb like *seem* does not subcategorize for an NP direct object but only for a C″. There is thus no object NP position available to *Peter*. The possibility of moving *Peter* to the ϕ position is realized through the Θ-theory. VPs assign a Θ-role to the subject position of their clause with the exception of those headed by so-called raising verbs, i.e. verbs like *to seem* that can have pleonastic subjects like *it* or *there*. On the other hand, the subject of *to have left* is Θ-marked. Now, Θ-theory includes the Θ-Criterion defined in (13) above which specifies that each argument must get one and only one Θ-role. Therefore, if *Peter* along with its Θ-role is moved to the φ position then the Θ-Criterion is not violated and the Case Filter is satisfied. In other words, a sentence like (16b) is generated through the processing of (16a) in the grammar. The active principles are, among others, Θ-theory and Case theory.

Now consider the examples in (17).

(17)a. φ to seem John to have seen Mary.
b. *John to seem to have seen Mary.
c. φ asks John to have seen Mary.
d. *John asks to have seen Mary.

In (17b) the Case Filter is violated since the original position as well as the landing site for *John* are not Case-marked positions. The sentence is blocked since a lexical NP is without Case. In (17d) *ask* is a verb the subject position of which is Θ-marked. Therefore if *John* is moved to this position it will have two Θ-roles: the one assigned by *to have left* and the one assigned by *ask*. This is a violation of the Θ-Criterion and the sentence is ruled out.

I turn now to a crucial part of the theory, the Projection Principle. The Projection Principle stipulates that the representations at each syntactic level are projected from the lexicon. In other words, the thematic properties of the lexical items must remain constant throughout the derivation; cf. Pesetsky (1982). If a verb, for example, is specified in the lexicon as taking an object NP, then, at each syntactic level, this object must be present. In (16c), for example, *make* has an object NP marked with the feature [+Wh]. After application of Move α resulting in (16d), *make* has no overt object. However, the Projection Principle is not violated since a trace is left in this position. We see therefore that the Projection Principle serves in the determination of the distribution of empty categories. I will also make use of the Extended Projection Principle introduced in Chomsky (1982). It specifies that subjects of clauses must be present (in particular the Spec of INFL″). Rothstein (1983) suggests that the "extended" part of the Projection Principle can be derived from the theory of predication developed in Williams (1980).

The empty categories dealt with in this book are all in NP position. More specifically, the problems under study here involve the pronominal empty categories PRO and pro. The former is the empty category found in subject position of infinitival clauses. Its reference is assigned by the theory of Control whereas its distribution is determined indirectly by the Binding theory. The fact that PRO has the feature specification [+pronominal, +anaphor] forces it to appear in ungoverned positions. Pro is the empty category associated with the subject position of tensed sentences in languages where subject-verb agreement is rich enough to determine its reference; cf. Chomsky (1982) for a discussion and 1.2 below. Contrary to PRO, pro is [+pronominal, −anaphor] which allows it to appear in governed positions. In fact, we will see that it has to be governed.

The empty category pro, its properties, reference, and distribution will be discussed in greater detail in 1.2.3 and in chapter 2.

1.2. NULL ARGUMENTS

Assuming the Extended Projection Principle seems to create a problem. In some languages, Italian, for example, it is possible to find sentences such as those in (18).

(18)a. Comprerò una macchina.
FUT + buy − 1s a car
'I will buy a car.'
b. Verrà.
FUT + come − 3s
'He will come.'
c. Lo conosce.
him + know − 3s
'He knows him.'
d. Ho telefonato.
have − 1s telephoned
'I telephoned.'

In these sentences, the subject is not present in the surface realization of the sentence. Obviously, this is impossible in non pro-drop languages, English for example, as can be seen from the morpheme by morpheme glosses in (18).

One's first reaction upon encountering such a property might be to say that it does not constitute a problem. One could simply assume that there is no subject position generated in D-structure in Italian in the sentences in (18). More precisely, the base structure rules produced by the X′ system will optionally generate a subject NP as sister of INFL′ as specified in (19). The sentences in (18) would have the structure in (20).

(19) C″ → Spec C C′
C′ → COMP INFL″
INFL″ → (NP) INFL′
INFL′ → INFL VP

(20)

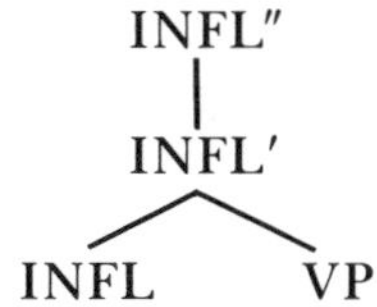

But there are major problems with this type of analysis. First of all, the Extended Projection Principle, which stipulates that the subject position is universally available, is violated within the framework adopted here. Secondly, in one possible analysis, the verbal inflection is obtained by a movement of the INFL node to VP (but cf. chapter 4). In its base position, INFL is in a certain relationship with the NP subject: INFL governs the subject. This relationship accounts for why the verb bears the particular inflection that is compatible with its subject, as in the French examples in (21).

(21)a. Nous voyons beaucoup de films.
we see-1P many of movies
'We see many movies.'
b. *Nous vois . . .
see-1S
c. Je vois . . .
I see-1S
d. *Je voyons . . .
I see-1P

In Italian the same process can be observed. A tensed verb always exhibits subject-verb agreement even when there is no (overt) NP subject, as in (18). This is another reason why a representation such as the one in (22) should be preferred over the one in (20). The structure in (22) allows us to generalize the process of subject-verb agreement in sentences with overt subject to the ones with no such subject.

(22)

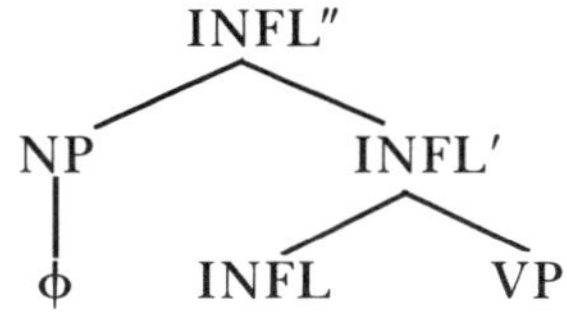

A third piece of evidence supporting the presence of a subject position can be drawn from binding theory. A lexical NP in object position as in (23) is considered to be an R-expression and, as such, falls under condition C of the Binding Conditions which stipulates that an R-expression must be free.

(23) $[_{I''}$ $[_{I'}$ INFL $[_{VP}$ vedrà $[_{NP}$ Giovanni$]]]]$

If one assumes the structure in (20) then the object NP in (23) is free and a possible interpretation for this sentence would be a reflexive reading corresponding to the reading in (24) where *he* and *John* are coreferent.

(24) *He$_i$ will see John$_i$.

But this interpretation of (23) is not available. The sentence has two possible readings. In the first, *Giovanni* is the subject of intransitive *vedere* after application of free inversion; see 2.4 for a discussion of this type of inversion. The sentence has the meaning in (25).

(25) Giovanni will see.

In the second interpretation, *Giovanni* is the object of *vedere* and is disjoint in reference with the subject. In that case the sentence means:

(26) He$_i$ will see Giovanni$_j$.

The structure in (22) allows us to properly rule out the reflexive interpretation in (24). This is so because the sentence has the representation given in (27).

(27) *$[_{I''}$ $[_{NP}$ $\phi_i]$ $[_{I'}$ INFL $[_{VP}$ vedrà $[_{NP}$ Giovanni$_i]]]]$

In (27), *Giovanni* is bound by the subject NP and therefore violates condition C of the Binding Conditions. On the other hand, the reading in (26) follows automatically since, under that reading, (23) has the structure in (28).

(28) $[_{I''}$ $[_{NP}$ $\phi_i]$ $[_{I'}$ INFL $[_{VP}$ vedrà $[_{NP}$ Giovanni$_j]]]]$

The same type of argumentation could be repeated for conditions A and B of the binding theory.

One obvious move to account for the ungrammaticality of the reading in (24) in a theory that does not accept empty subjects as in (27) would be to assign a referential index to AGR in INFL and assume that this element occupies an A-position. The resulting structure is given in (29).

(29) *[$_{I''}$ [$_{I'}$ [$_{INFL}$ AGR$_i$ TNS] [$_{VP}$ vedrà [$_{NP}$ Giovanni$_i$]]]]

In (29), *Giovanni* is bound in violation of condition C and the coreferential reading is excluded. If AGR and *Giovanni* bear different referential indices then the right interpretation is obtained.

One problem with such an account is that certain languages such as Chinese and Japanese exhibit null subjects without any overt subject-verb agreement; cf. Huang (1982;1984) and 1.2.2 for more on this type of empty argument. Unless empty AGRs are assumed, the solution in (29) with AGR bearing a referential index is not available and the Binding Conditions violations are unaccounted for.

We would then want to ask why a theory allows empty AGRs while rejecting empty NPs in order to account for a given phenomenon. I thus assume the theory that allows for structures like the one in (22) as the representation for tensed clauses with empty subjects in languages like Italian.

1.2.1. Previous Analyses

Now that we have a basic statement of the problem, we want to ask what makes it possible to have null subjects in tensed sentences in some languages and not in others? This section explores the different attempts that have been made at answering this question. Recalling the discussion in 1.1, the issue here falls under the type of analysis involving cross-linguistic variation. The possibility of allowing empty subjects should therefore be parameterized and learnable. Once the parameter(s) are set, we want to ask whether any other syntactic properties can follow from the account.

The study of the set of properties associated with null subject languages, that is, the syntactic characteristics of these languages that can be said to follow from the fact that the subject can be empty, was initiated in Perlmutter (1971). His generalization, approximately stated, says that null subject languages do not exhibit the usual "that-trace" effects. Compare English to Italian in (30).

(30)a. *Who did you say that came?
b. Chi credi che partirà?
'Who do you think that will leave?'

In order to explain this characteristic, it was first assumed that, in languages allowing both (30b) and empty subjects, an optional rule of subject deletion could delete the trace left by extraction of the subject as well as a lexical pronoun; cf. Chomsky and Lasnik (1977). Null subject languages are thus referred to as "pro-drop" languages.

The next step attempted to relate the missing subject property to a rich verbal inflectional system. Taraldsen (1978) points out that English has a very poor subject-verb agreement system whereas Italian has one which is, by comparison, quite complex. Following this observation, Chomsky (1981) proposes that the inflectional element INFL contains the agreement element AGR which has some abstract property that makes it possible to recover the information made unavailable by the fact that the subject is missing; see 2.2.2 for more on agreement. English differs from Italian in that the latter exhibits this property whereas the former does not. Chomsky discusses a set of properties associated with the null subject parameter. They are given in (31) along with the relevant examples from Italian.

(31)a. Missing subject:
Ho trovato il libro.
'(I) found the book.'
b. Free inversion in simple sentences:
Ha mangiato Giovanni.
'Giovanni ate.'
c. Long *wh*-movement of subject:
L'uomo [che mi domando [chi abbia visto]].
'The man (x) such that I wonder who x saw.'
d. Empty resumptive pronouns in an embedded clause:
Ecco la ragazza [che mi domando [chi credi [che possa VP]]].
'This is the girl who I wonder who thinks that she may VP.'
e. Apparent violations of the that-t filter:
Chi credi [che partirà]?
'Who do you think (that) will leave?'
(Chomsky 1981, 240)

Chomsky proposes a solution based on Rizzi's (1982) work. Rizzi argues that extraction of the subject in Italian takes place from a governed postverbal position and therefore does not violate the ECP at LF. This solution consists, in fact, of reducing the cluster of properties peculiar to null subject languages in (31) to the ones in (a) and (b), namely missing subject and free inversion. Since no extraction operation is involved in these constructions, Chomsky assumes that the empty subject position is PRO cosuperscripted with AGR. But

PRO, being a pronominal anaphor, is excluded by the Binding Conditions from having a governing category, thus deriving the theorem that PRO must be ungoverned at the level where the Binding conditions apply, that is, at S-structure. Assuming that for all languages with subject-verb agreement, a rule moves INFL inside VP (call it rule R), and that INFL is a governor for the subject position in tensed clauses, it is possible to propose the parameter in (32).

(32) R may apply in syntax.

(Chomsky 1981)

In a null subject language this option is available and, if taken, PRO appears ungoverned by INFL at S-structure making it possible to have an empty subject and to assign nominative Case to the inverted subject dominated by VP in the instances of free inversion; cf. also Borer (1984). In English, for instance, in which (32) is fixed negatively, PRO in subject position of a tensed clause will always be governed at S-structure – since R applies in PF – and is therefore excluded from this position. Furthermore, free inversion is blocked since Case cannot be assigned to the inverted NP by INFL. Note that in the instances of free inversion, the empty category in subject position is pleonastic PRO analogous to English *there* and French *il*.

In his discussion of the typology of empty categories, Chomsky (1982) introduces the empty category pro corresponding to an overt pronoun, i.e. a pronominal non-anaphor, and argues that it is best suited to be the missing subject in tensed clauses in null subject languages. One of the reasons underlying this assumption is that a missing subject in these languages cannot normally be arbitrary in reference. This property is not accounted for if PRO is used in this position since PRO may be arbitrary. Furthermore, in Spanish interrogatives there is an obligatory verb-fronting operation which places the subject in a position (properly) governed by the fronted verb; see Torrego (1984). This makes it impossible to use PRO in subject position for the reasons stated above.

The use of pro has the further consequence of making it unnecessary to invoke a parameter such as the one in (32) since pro, as opposed to PRO, can be governed at S-structure. What must be added to the theory is the requirement that pro be identified or locally determined. The AGR element is rich enough in Italian that it can act as an identifier for pro in [NP,IP] whereas this is not the case in, say, English. In conclusion, pro can only appear in subject position in languages with rich subject-verb agreement.

One of the goals of Safir (1985) is to challenge any approach to the missing subject problem in which the absence of a subject and free inversion are closely related, as in Chomsky (1981) and Jaeggli (1982). His arguments are based on the existence of two dialects of Italian, Trentino and Modenese, that are not null subject languages but which have free inversion, and Portuguese, which exhibits the missing subject property without any possibility of free inversion. Free inversion is made possible in Safir's analysis by the assignment of nominative Case by a subject clitic to a postverbal NP. This causes some difficulties for any treatment of the null subject parameter that relates it to free inversion.

In Bouchard (1982), the null subject property is parameterized through the option of delaying the percolation of the Case assigning feature of INFL until LF. In this analysis, INFL is in VP in syntax (D-structure, S-structure, and LF). According to his Principle of Lexicalization, which stipulates that a nominal is lexical if and only if it has certain features at PF (person, number, gender, and Case), if a nominal is assigned Case only at LF by INFL in VP, it will be "invisible" at PF. That is the case in null subject languages. On the other hand, non-null subject languages require the Case-assigning feature of INFL to percolate to the VP projection before or at S-structure. The nominal in subject position is therefore always Case-marked and it follows that it will be lexicalized in PF.

To conclude this section it should be noted that all the analyses presented here, except for Bouchard's discussion of Japanese, concentrate on Romance languages where we often find rich subject-verb agreement. Two types of languages are neglected: those with no agreement at all and those with object-verb agreement. The following section reviews Huang's (1982;1984) treatment of the former type. I briefly discuss the latter type in chapter 4.

1.2.2. Zero Topic Languages

Let us take (Mandarin) Chinese as an example of a zero topic language using Huang's (1982;1984) terminology and data. Chinese has no agreement at all as can be seen in (33) and (34). I omit the tones.

(33)a. Wo shi xuesheng.
I be student
'I am a student.'

b. Ni shi xuesheng.
you(s) be student
'You are a student.'

c. Ta shi xuesheng.
he/she be student
'he/she is a student'

(34)a. Women lai
we come
'We come.'
b. Nimen lai.
you(p) come
'You come.'
c. Tamen lai.
they come
'They come.'

The form of the verb is invariable and this can be analyzed as implying that there is no AGR in INFL in Chinese. If this is true then finite clauses in Chinese should behave like infinitives in English with respect to the content of the external argument position (the subject).[6] In other words, it should be possible to find an empty subject in finite clauses in Chinese. The examples in (35), from Huang (1982, 352) show that this is indeed the case.

(35)a. Zhangsan shuo [[ϕ] xiawu hui lai].
say afternoon will come
'Zhangsan said that he will come this afternoon.'
b. Zhangsan shuo [[[ϕ] you-mei-you qian] meiyou guanxi].
have-not-have money no matter
'Zhangsan said it doesn't matter whether he has money or not.'

Huang reports that the empty subject of the lower clause in (35) is interpreted as referring to *Zhangsan*. Huang assumes that it is pro, the empty pronominal non-anaphor, and that it must be coindexed with the closest nominal element; see Huang (1982, 365ff). This is reminiscent of control theory which assigns a reference to PRO. Note here that, in a certain sense, English can be considered to exhibit the null subject property in a very restricted manner. In effect, an empty subject appears in infinitival clauses in this language as in (36).

(36) John wants [[ϕ] to go].

This leads Huang to observe that null subject arguments appear in two circumstances: when there is very rich agreement and when there is no agreement at all.

In the second chapter of this book we will see that languages with subject clitics must be analyzed as null subject languages. Whether they must be considered to be of the rich agreement type will be considered.

The difference between (35) and (36) and the other Chinese sentence in (37) with respect to identification of the empty subject is that, in (37), the empty category is not identified by the principles of control.

(37) [ϕ] lai le.
come ASP(ect)
'He came.'

(Huang 1982, 355)

Rather, it is assumed that an empty operator which Ā-binds the empty category is present in a position adjoined to IP as in (38).

(38) $[op_i] \ldots [_{IP} [\phi_i \ldots]]$

This assumption is supported by the fact that the empty category discussed here can appear in object position as well as in subject position (37). Thus, the discourse in (39) is possible in Chinese.

(39)a. Q: Lisi$_i$, shei kanjian-le t_i?
who see-ASP
'As for Lisi, who saw (him)?'
b. A: Zhangsan shuo [ni kanjian-le [ϕ]].
say you see-ASP
'Zhangsan said that you saw (him).'

(Huang 1982, 355)

What (39b) shows is that the empty category can refer back to a topic (*Lisi*) introduced in a previous discourse (39a), even though this topic is not phonologically present in the sentence (39b). This topic is called a "discourse-topic" and is represented structurally as (38). This is the characteristic of the zero topic languages that is relevant here; the ability to refer to a discourse-topic is part of the reason it is possible to have empty arguments of the type found in (37) and (39b) in these languages.

Finally it is important to note that the type of empty subject discussed here is different from the one in tensed clauses in non zero topic languages like Italian. Consider again the sentence in (37) above. Contrary to the Italian counterparts in (18), the empty category in (37) cannot be pronominal since there is no "closest" nominal element (there is no AGR). Huang therefore proposes that it is a variable Ā-bound by a discourse topic as in (38); but cf. Liejiong (1986) for a different approach.

According to Huang's analysis, the null subject property as it is found in Italian is not incompatible with the zero topic possibility. The two are different and interact in particular ways in languages. Italian, for instance, is a non zero topic, null subject language whereas English is non zero topic, non-null subject. Chinese would be zero topic (37) and null subject (35). German, in this view, could be analyzed as zero topic but non null subject; see the end of section 2.2.1 for more on German.

The discussion in this section forces the conclusion that, in order to achieve a higher level of explanation, any analysis of the null argument property must include zero topic languages such as Chinese and Japanese. Whereas the present research concentrates on problems related to the type of null arguments found in languages like Italian and languages with clitics, we include the zero topic languages in the larger typology of the languages allowing null arguments because they are all applications of the principle of recoverabilty discussed in the introduction.

1.2.3. A Theory of pro

This section is crucial for the development of the rest of this book. As we have seen, Chomsky (1982) introduces, for various reasons, the empty category corresponding to a pure pronominal, that is, pro. The problem is that the conditions on the appearance of pro are not defined precisely enough. The following questions arise: (A) What kind of element is sufficient for its licensing? and (B) Are the licensing of pro and the assignment of content to pro two distinct operations? and if not, why does pro differ from the other types of empty categories in this respect? For instance, the Extended Projection Principle allows PRO (an NP) to appear in argument position. This position must be an ungoverned one according to the binding theory. PRO is assigned its feature content by the theory of control. In the case of pro, both its presence in the structure and its particular feature-content are made possible by government by rich inflection.

The second question is treated in this section; I return to the first in chapter 2. The discussion here is based on Rizzi (1986b) who first raised question B.

Identification of pro, as shown in 1.2.1, is done through rich inflection. Formally speaking, when the element AGR of INFL has a certain property (that of being rich), it can assign its features to pro under government. The problem pointed out in Rizzi (1986b) concerns the appearance of pro in contexts where this condition on its identification is not met. Consider the Italian examples in (40) from Rizzi.

(40)a. Questo conduce {la gente / ϕ} alla seguente conclusione.
'This leads (people) to the following conclusion.'

b. Gianni è sempre pronto ad accontentare {la gente / ϕ}.
'Gianni is always ready to please (people).

According to Rizzi, in the cases where the object is empty, the internal argument position can only be filled by pro.[7] But then the problem arises of properly identifying this element. Rizzi proposes the following licensing principle:

(41) pro is Case-marked by X_y.

(Rizzi 1986b, 524)

where X is a governing head of the type y. Variation among languages is accounted for by the different values assumed by X_y in a given language. In English, X_y has no value and pro can consequently never be licensed in any structural position. In Italian, X_y can be INFL for pro in subject position and V for pro in object position.

The content of pro is determined through the following convention:

(42) Let X be the licensing head of an occurrence of pro; then pro has the grammatical feature specification of the features on X coindexed with it.

(Rizzi 1986b, 520)

To see how this convention operates, let us examine the Italian sentences in (43)-(45).

(43)a. Verrà.
FUT + come − 3S
'He will come.'
b. [$_{IP}$ pro$_i$ INFL$_i$ verra]

(44)a. Si dorme troppo.
cl sleep-3s too + much
'People sleep too much.'
b. [$_{IP}$ pro$_{arb}$ [$_{INFL}$ si$_{arb}$] dorme troppo]

(45)a. La buona musica riconcilia ϕ con se stessi.
'Good music reconciles ϕ with oneself.'
b. la buona musica [$_{VP}$ [$_V$ riconcilia] pro$_{arb}$ con . . .]
[Θ_{arb}]

In (43) pro is licensed by the nominative assigning head INFL and its content is determined by the same element. In (44) pro is licensed by INFL but its features are specified by the arbitrary subject clitic *si* under INFL. In (45) pro is licensed by the Case assigning head V and its content is defined through coindexing with a slot in the Θ-grid of V (analogous to coindexing with AGR under INFL). The internal Θ-role corresponding to the object position is assigned the index *arb* in the syntax and, according to the Projection Principle, this index must be transferred to the NP object; see Rizzi (1986b) for further details.

This is essentially the theory of pro developed in Rizzi (1986b). In chapter 2 we will see whether this theory is necessary and sufficient to answer question A above.

1.3. ON EMPTY CATEGORIES

Returning to the discussion in the Introduction, it seems clear that a gap in a sentence must be recoverable in order for communication to be successful. It is therefore plausible to assume that particular grammars, for independent reasons, allow certain elements in certain contexts to be missing. We have seen in the previous sections that these gaps are represented as empty categories in the structural representation of the sentence. We have also seen under what circumstances null arguments like pro and PRO in Italian and Chinese, for example, are allowed to appear in structural positions.

Besides pro and PRO, it is generally assumed in the Government and Binding framework that NPs can be represented by two other empty categories, namely variables and NP-traces. In this section, I will briefly enter an area of controversy in the field concerning the determination of empty categories in order to make clear my assumptions concerning them.

The first problem can be expressed as the following simple question: How are the different types of empty categories defined in the grammar?

The approach adopted here is as follows. Some empty categories are base generated (pro and PRO) and some are produced by movement of NPs in the transformational component, LF or possibly PF (NP-traces and some variables).

A gap in a structure receives its features from the structural context and the various features that are available come from binding theory.

One of the consequences of this approach is found in the application of movement from an NP position that is governed by rich agreement. The D-structure of such a sentence has the form in (46).

(46) $[_{CP} [_{IP} [_{NP}\mathit{wh}] [_{I'}$ INFL VP]]]

The S-structure is as the one in (47).

(47) $[_{CP}\ \mathit{wh}_i\ [_{IP} [_{NP}\ e_i] [_{I'}$ INFL VP]]]

The problem consists of determining what type of empty category occupies the NP position in (47) from where movement originates. Recall that we assume that INFL contains AGR with all the necessary features for licensing pro. So, is *e* in (47) pro? It can be if we assume that pro is governed by rich agreement. On the other hand, the empty category is $\bar{A}$-bound by the *wh*-antecedent in Spec COMP and such an empty category is a variable if we assume that variables are $\bar{A}$-bound. Thus, in this particular case, the context allows for two different types of empty categories: pro and variable.[8] Pro is not excluded by definition since the Binding conditions only stipulate that it must be A-free. In (47) the empty category is A-free and could therefore be pro. This is an important problem that must be resolved since the context allows two interpretations of the gap and neither of them is "intrinsically better" than the other. I will assume in the rest of this study that *e* in (47) is a variable produced by movement and that it cannot be pro, which is a base generated empty category. In this view, pro can act as a variable only if it is base-generated as such; cf. 3.3.3 for a direct application of this, and Jaeggli (1984) for a different view.

Note that there is at least one consequence of this assumption, which is that we can maintain the distinction between empty resumptive pronouns and variables. Suppose that a language allows *wh*-movement. Suppose further that it has rich agreement. Assume finally that we took the opposite view, namely that an empty category governed by rich agreement is always pro even if movement took place from the position it occupies. Then an empty category in subject position in this language is always pro and, when coindexed with an operator, it is a resumptive pronoun. Consequently, there will never be a variable in subject position in any of the derivations possible in this language and the useful distinction with respect to Subjacency between pro coindexed with an operator and trace coindexed with an operator will be lost. This is not a desirable consequence.

I thus assume that movement operations leave traces in the syntactic structure. The reader is referred to Chomsky (1981;1982), Bouchard (1982), and Brody (1984) for extensive discussions of the problem of defining the empty categories.

The second issue has to do with the reference to empty categories with respect to UG. The question is, Should the principles, rules, parameters, and filters of UG make explicit reference to empty categories?

Two views can be proposed. According to the first, which is widely held in current studies in the generative framework, there is no theoretical or metatheoretical objection to any explicit reference to empty categories. The null subject parameter, for example, can therefore make reference in its formulation to pro and, informally, state that pro is licensed by rich inflection or a given category. Hence only a language exhibiting rich inflection can be a null subject language; cf. 1.2. The other view, defended especially in Bouchard (1982), is based on the assumption that there is only one empty category that exhaustively partitions the distribution of lexical NPs. The distribution, type, and reference of lexical NPs are determined by the interaction of other independent components of UG. Therefore, the distribution, type, and reference of empty categories – which are NPs – should also derive from the same interactions. It follows that no principle, rule, parameter or filter should make reference exclusively to empty categories since they are regular NPs. In other words, the components of UG do not discriminate between empty and overt NPs; cf. Bouchard (1982) for the development of such an approach and an examination of its consequences.

I adopt here the latter view because it appears to provide the more restrictive model. The consequences of this choice will be visible throughout the rest of this book and will guide decisions to be made when confronted with more than one plausible analysis. One example of this type is illustrated by my attempt, in chapter 3, to formalize the relationship existing between clitics and related argument positions. This formalization should make no reference to whether or not the argument position must be empty or phonologically realized. The only relevant matter is whether or not the NP occupying it can be bound or not. The problem of determining the empty or overt status of the NP is a separate issue addressed in chapter 2. The particular status must follow from whether the position occupied by the NP is Case-marked or not.

To conclude, whenever possible, I will avoid explicit reference to the status of the NPs with regard to their being empty or overt in the parameters and definitions proposed in this book.

1.4. SUMMARY

We have seen the particular approach to the language faculty adopted here, and how, by its use of principles and parameters, this approach can account for variation among languages and dialects on one hand, and for language growth on the other hand. The relationship between variation and language development in the individual is assumed to be crucial because it restricts the way the various problems related to the syntactic recoverability of null arguments can be approached. A parameter must be formulated in such a way that plausible trigger experiences for its proper settings in core grammar are available to the language learner.

The representation of UG assumed here was then presented together with the principles it is composed of. A core grammar is thus conceived of as the interaction between those principles, different parameters, and the subcategorization properties of the lexical items within the representation in (2).

I then turned to the more specific problems that are discussed in the following chapters, namely the syntactic recoverability of null arguments as instantiated in null subject languages. Some of the previous analyses of this property were reviewed. It was pointed out that the specific type of null argument found in zero topic languages was neglected in most of those analyses. In 1.2.2, Huang's analysis of the zero topic languages was introduced and this was followed by a brief discussion of the interaction between this property and the null subject property as found in, say, Italian. The theory of pro developed in Rizzi (1986b) and which will be further tested in the following chapters was summarized in 1.2.3.

CHAPTER TWO

Null Arguments in Romance Languages

2.0 INTRODUCTION

The main purpose of this chapter is to present an analysis of subject clitics in which they are base-generated under AGR of INFL.[1] After motivating this approach on the basis of cross-linguistic observations in 2.1, I will argue that the only possible filler for the subject position is a phonetically realized NP and depending on the language, pro. As we have seen in the first chapter, it is now generally accepted that null subject languages are the ones in which the external argument position can be occupied by pro. This leads to the proposal that French as well as the other languages with subject clitics can be considered to be null subject languages in which pro in [NP,IP] is licensed by a clitic and that the grammatical notion "subject clitic" is not a spurious one. It is argued in Rizzi (1986a) that the elements generally referred to as "subject clitics" can be generated in different syntactic positions across languages; see 2.2.4 for some details. I wish to argue against this conclusion and assume that subject clitics in the languages under study here are always generated in INFL. The details of how the licensing of pro by a clitic operates are presented in 2.2, especially with respect to Case and Θ-role.

The discussions in this chapter lead to the conclusion that there is no pro-drop or null subject parameter per se. The possibility that a language exhibits null arguments follows from other subtheories of UG such that no explicit reference to this property is needed in any particular grammar. Section 2.3 puts this result in perspective by comparing it to the treatment of this property in Chomsky (1981;1982). I then continue with a look at two constructions that seem to cause problems for the general analysis in the preceding sections, namely free inversion in Standard Italian and some of its dialects, and causatives in Standard French.

2.1 A GENERALIZATION

Rizzi (1986a) appeals to some distributional facts concerning subject clitics in certain Northern Italian (N.It.) dialects and Standard French (St.Fr.) to argue that the notion "subject clitic" defines a spurious syntactic class.[2] This conclusion is based on the assumption that the N. It. dialects represent an intermediate state between Standard Italian (St.It.) and St.Fr. In the N.It. dialects a preverbal subject clitic is obligatorily present even when there is a phonetically realized preverbal subject NP. Rizzi suggests that such clitics are phonetic realizations of the AGR node in INFL. These N.It. dialects would represent intermediate states between St.Fr. and St.It. in that, as in St.Fr. and contrary to St.It., they have subject clitics but, as in St.It. and contrary to St.Fr., they also allow the empty pronominal pro in subject position. So, St.Fr. and the N.It. dialects would differ by the fact that subject clitics in the latter license pro in subject position whereas they cannot do so in the former; see 2.2.4 for more on Rizzi's (1986a) analysis. The properties of the subject doubling phenomenon found in the N.It. dialects will be discussed extensively in chapter 3. For now I will simply assume as a working hypothesis that the notion "subject clitic" is a grammatically real one. This hypothesis is based on the existence of another apparently intermediate case between St.Fr. and the N.It. dialects, represented by Colloquial French (C.Fr.) and Pied Noir French (PN Fr.).[3] In C.Fr. there also exists a possibility of subject doubling which differs from that of the N.It. dialects in that it is optional. I thus assume that subject clitics in C.Fr. and in the N.It. dialects must be generated in the same syntactic position. This analysis is then extended to St.Fr.

Thus, in some languages, it is possible to find sentences where a lexical NP subject is doubled by a coreferent clitic pronoun. These pronouns are called "clitics" because they are affixed to the inflected verb in the same sentence at least on the surface. The doubling construction can be represented schematically as in (1), where SCL stands for subject clitic.

(1) NP SCL V . . .

A plausible account of this construction which I will assume consists in saying that, contrary to pronouns in English, clitic pronouns are not generated in subject position. This position is therefore free to receive a lexical NP under certain circumstances to be examined in 3.1. But then we can ask what occupies the subject position when it is not filled by lexical material as in (2).

(2) $[_{NP}\ \phi]$ SCL V . . .

As we will see, this gap seems to meet all and only the conditions necessary for the presence of the pure empty pronominal pro. This is basically the approach assumed in Rizzi (1986a) for the N.It. dialects. In this chapter, I will be concerned with the representation in (2) and will delay discussion of (1) to the next chapter. In particular, I want to argue that the representation in (2) must be valid for all languages with subject clitics.

To do so, let us postulate the existence of two groups of languages and dialects: those with subject clitics and those without. In the latter group, subject pronouns are found in the external argument position [NP,IP]. The first group includes, for example, St.Fr. and C.Fr., the N.It. dialects (including Trentino, Fiorentino, Modenese), Pirahã, and Pashto[4]. In the second group are languages such as Italian, Spanish, English, German, and Modern Hebrew.

It is a well known fact that some of the languages of the group with no subject clitics are so-called pro drop languages (Italian, Spanish) whereas others are not (English, German).[5] Interestingly, there seems to be no known member of the group with subject clitics that is a pro drop language in which there is no overt subject at all be it a full lexical NP or a pronoun (clitic or not).[6] Given the representation in (2) above, this observation can be stated as the generalization in (3).

(3) In a language with subject clitics, the [NP, IP] and SCL positions cannot both be empty in simple tensed clauses.

In fact (3) states that the representation in (4) is barred in languages with subject clitics.

(4) $[_{NP}\ \phi]$ $[_{SCL}\ \phi]$ V

The first version of the generalization makes explicit reference to "simple tensed clauses" because there are various constructions where it does not hold. In (5), for instance, after extraction of the subject both positions are empty in French as in (4).

(5)a. Qui_i $[_{IP}\ t_i$ va venir]?
'who will come?'
b. *Qui_i $[_{IP}\ t_i\ il_i$ va venir]?

The same is true for French Stylistic Inversion where the subject is realized postverbally with no overt material in preverbal position.

Such derived constructions are not directly relevant here. The point is that a simple tensed clause in a language with subject clitics cannot have the form in (4). This has to be accounted for.

Turning to the second group (with no subject clitics), it is generally assumed, as we have seen in 1.2, that a property of INFL, which is probably rich subject-verb agreement, is sufficient to allow the NP subject to be occupied by pro. As we have seen in Chapter 1, the agreement is said to be "rich" when the verbal morphology carries agreement features that are lexically realized in such a way that this morphology is generally different for each person and number. The difference between, say, English, which does not have rich agreement, and Italian, which does have rich agreement, is thus accounted for.

Let us assume that, in the first group, the subject clitic is just another manifestation of a property of INFL allowing pro in [NP,IP]. I will return shortly to the details of how this is done. If this is true, then the number of null subject languages is much larger than previously thought since it includes the ones with subject clitics as well as those with rich subject-verb agreement or no agreement whatsoever (see 1.2.2).

We can now reformulate the generalization in (3) in more formal terms in order to eliminate possible ambiguities. The "simple tensed clauses" restriction is replaced by the mention of pro. The empty category in [NP,IP] after extraction of the subject as in (5a) is a trace which does not fall under the generalization.

(6) If the grammar of L includes subject clitic pronouns then, when [NP,IP] is phonetically null and filled by pro, a subject clitic must be present.[7]

One of the numerous questions raised by the preceding discussion concerns the status of this generalization in UG. I want to argue that it can be derived from other principles of grammar.

2.2 PRO AND CLITICS

In this section I present an implementation of the generalization in (6). To put it differently, I account for why a clitic and/or a lexical NP must appear in tensed clauses in languages with subject clitics. The analysis to be presented is supported by the syntactic similarities existing between object clitics and subject clitics in the non-doubling cases. The asymmetries between subject doubling and object doubling will be shown to derive from other principles in the next chapter. I believe that an account in which subject clitics and object clitics are simply different structural realizations of the same unique notion "clitic" is preferable to one in which they are quite different syntactic elements, both for the sake of a coherent theory of syntactic recoverability and for learnability reasons.

2.2.1. On Pro

As we have seen, Chomsky (1982) presents a typology of empty categories where pro, corresponding to the null realization of the features [+pronominal, −anaphor], is identified strictly by the element AGR of INFL and only when this element is rich enough. I assume this basic idea in the rest of this book with the following points of elaboration. Pro must be licensed by a set of features usually associated with nominal elements and agreement markers. These features are [person, number, (gender)]. Contrary to the proposal in Chomsky (1982) and Rizzi (1986b), I do not take [Case] to be a member of the set of features required for the licensing of pro. Case-assignment is therefore not the licensing domain for pro in the present approach as it is in Rizzi (1986b). The reasons for this assumption will become clear in the following sections, in particular 2.2.5. I assume further that these features must be, in some obvious sense, lexically realized. There are two reasons for this assumption. First, it is natural to suppose that in languages where there is no rich subject-verb agreement, in English for example, each realization of a finite verb bears the required features even though they are not lexically realized. These features on the verb cannot license pro since they are not lexically realized. Secondly, this condition on the lexical realization of the licensing features must be adopted for learnability reasons. How would the language-learner be able to analyze its language as involving the type of recoverability under discussion here if nothing on the surface allows for this recoverability? The set of phonetically realized features must also govern in a very local sense and be coindexed with the argument position occupied by pro. Finally, I assume that there is no such thing as a licensing domain for pro where this notion is understood as one imposing restrictions on where pro can appear at D-structure. In conformity with the discussion on empty categories in 1.3, I maintain the idea that pro, being a pronominal element with no phonological matrix, that is, an empty pronoun, should not have restrictions imposed on it as far as its position is concerned. In other words, pro is allowed in all and only the positions where an overt pronoun can appear at D-structure. These positions are: Θ-marked A-position, non Θ-marked A-position, and $\bar{A}$-position.

To summarize the assumptions with respect to pro, I have said:[8]

> Pro can appear anywhere an overt pronoun is allowed and it is "licensed" in these positions if it is coindexed with and governed by some (phonetically) realized set of features including person, number and, in some cases, gender.

It follows that any particular grammar has the potential to license pro; variation among languages lies therefore in the fulfillment or non-fulfillment of the condition on the recoverability of the features of pro. I return to the consequences of this in 2.3.

I now briefly review some of the putative syntactic positions of pro that are attested in the literature in order to illustrate the effect of the above condition on this empty category.

One of the positions available for pro is as object of a verb. Most of the cases studied in the literature involve various object clitics in Italian (Rizzi 1982), Spanish (Jaeggli 1986). French (Zubizarreta 1982b), Arabic (Aoun 1979), and Romanian (Farkas 1978). Some examples are provided here.

(7) Italian
Gianni la_i presenterà pro_i a Francesco.
her introduce-3s to
'Gianni will introduce her to Francesco.'

(8) Spanish
Juan lo_i visito pro_i.
him visited-3s
'Juan visited him.'

(9) French
Marie la_i mange pro_i.
it eat-3s
'Marie eats it.'

(10) Arabic
Šif -t $-a_i$ pro_i.
saw -1s -him
'I saw him.'

(11) Romanian
L_i -am văzut pro_i.
him -have-1s seen
'I have seen him.'

In all the preceding examples, the two conditions on the appearance of pro are satisfied: the verb Θ-marks its (direct) object position such that it could be occupied by a pronoun and the object clitic affixed to the verb bears the necessary features for proper "licensing." Note that these examples are ruled out in these languages if the clitic is not present and if the null object is not the result of extraction.

(12)a. *Gianni presenterà pro a Francesco.
b. *Juan visito pro.
c. *Marie mange pro.
d. *Šif-it pro.
e. *Am văzut pro.

Thus, even though pro as a pronoun could appear in these positions it is not licensed by the appropriate features. In other words, the content of pro is not recoverable.

The next context where pro is attested is as object of a preposition. Stowell (1981, 144), following a suggestion of Hagit Borer, proposes that PP complements themselves are not Θ-marked but that it is the NP complement of the preposition which receives a Θ-role assigned compositionally by the verb and the preposition. Whether this is true or not, it is natural to assume that pro can appear in this position as long as it is licensed. In the following examples from Arabic and Modern Hebrew, a clitic appears on the preposition and carries the necessary features for the licensing of pro; cf. Borer (1984).

(13)a. Modern Hebrew
'it $-o_i$ pro_i
with-him
b. Arabic
$wiya\text{-}ha_i$ pro_i
with-her

Here again the above examples are barred if the clitic is absent.[9]

I turn now to the so-called construct state in Arabic and Modern Hebrew. This construction expresses genitival relations between a head noun and a complement noun. As in the examples in (12) and

(13), we can assume that the complement position in the construct state is one potentially available for pro since an overt pronoun is allowed. Nevertheless, pro should appear only if a particular grammar allows the necessary lexicalized features to appear on the head noun. This is the case in the examples below where a clitic is affixed to the head noun; cf. Borer (1984).

(14) Modern Hebrew
a. beit -o_i pro_i
house-his
b. beit -a_i pro_i
house-her

(15) Arabic
ʔumm -a_i pro_i
mother-his

If the clitic is omitted, the NPs in (14) and (15) are ruled out.[10]

The subject position in different sentences from different languages poses some interesting problems. This follows from the fact that the external argument position of certain verbs does not receive a Θ-role. In this case a referential subject is possible only if it is moved to this position from a Θ-marked position; cf. 1.1.1.3. I will first look at some licit Θ-marked external argument positions for pro and will then explore what happens in the case of expletive subject positions.

As we have seen in 1.2, Spanish and Italian among others allow empty subjects in tensed clauses.

(16) Spanish (Jaeggli 1982)
a. pro baila bien.
dance-3s well
'He/she dances well.'
b. pro estamos cansadísimos.
be-1P very + tired
'We are very tired.'

(17) Italian (Rizzi 1982)
a. pro comprerò una macchina.
buy-1s one car
'I will buy a car.'
b. pro verra.
come-3s
'He will come.'

Recall that in these examples the external argument position is Θ-marked and can contain a full pronoun but that it is not obvious how pro is licensed. This is where the rich inflection property of the verbal paradigms in Italian and Spanish is important. It is usually assumed that pro is governed by and coindexed with AGR of the head INFL of the sentence as in (18).

(18) $[_{IP} [_{NP} pro_i] [_{INFL'} [_{INFL} AGR_i\ TNS] [_{VP} \ldots]]]$

The AGR element of INFL contains the necessary features; pro is thus licensed. I return to the possibility of having a subject clitic as licenser of pro in external argument position in 2.2.2.

The cases where the external argument does not receive a Θ-role directly from the VP are less straightforward. Consider the following Italian sentences.

(19) Non è chiaro come raccontare certe storie.
'(It) is unclear how to tell certain stories.'

Let us first see how this type of construction works in languages with overt pronominals. In English the string surfaces with the overt pleonastic pronoun *there* or *it* as in (20).

(20)a. It seems that John left.
b. There is a man in the garden.

These pleonastic pronouns are not referential and are only allowed in non Θ-marked subject positions. Their presence is obligatory since the external argument position, even though it is not projected from the lexicon by the Θ-marking properties of the verb, is nevertheless required by the Extended Projection Principle. Alternatively, it can be assumed that the pronouns are lexical realizations of nominative Case. That is why these pronouns are expletive; they have no interpretation with respect to thematic relations. Now, they cannot be phonologically null since pro is not licensed in English given the poverty of verbal inflections; but cf. the introduction concerning particular English cases and 2.2.2 on verbal inflections. The same analysis applies to languages with rich inflection in which the subject can be null. Even though the external argument position is not Θ-marked, it is nevertheless obligatory due to the Extended Projection Principle. Once this is established, it must be assumed that an expletive pronoun will appear in the subject position. Now, in null subject languages, pro can be licensed. It follows that the subject of a verb that does not have an external Θ-role to assign will be occupied by a non-referential (non-thematic) expletive pro. Therefore, given the Extended Projection Principle and the condition on proper licensing of pro, it is predicted that expletive pro exists and occupies non Θ-marked external argument positions in languages where its nominal features can be recovered. In other words, the Extended Projection Principle is, in a certain sense, sufficient to guarantee the presence of pro, which is referential when thematic relations are involved and expletive otherwise.

Notice that this discussion cannot be extended to non-subcategorized object positions. These positions are not Θ-marked but their occurrence violates the Extended Projection Principle since they are lexically governed but not projected from the lexicon.

These remarks have implications for the analysis of inversion facts in languages like Italian and Spanish. I examine them in 2.4.

The last construction that I will discuss here is the subject position of infinitives. In languages like French and English, such a subject is generally phonologically null since it is not governed and, therefore, does not get Case from the TNS element of INFL.

(21)a.	*J'espère	Jean	boire	du	café.
	I hope-1s		to + drink	of + the	coffee
	'I hope Jean to drink coffee.'				
b.	J'espère boire du café.				
	'I hope to drink coffee.'				

(22)a. *I hope John to drink coffee.
b. I hope to drink coffee.

A lexically realized subject will be allowed in English and other languages only if there is an "exceptional" means for it to be Case-marked; cf. Massam (1985) for more on exceptional Case-marking. Because this position is usually ungoverned, it can be filled by PRO, the pronominal anaphor. What is more relevant here is that the external argument position of non tensed clauses with non raising verbs is Θ-marked and thus constitutes a potential position in which pro could appear. Two possibilities arise. In the first, the particular grammar of the language does not include any agreement process between the subject and the verb. Pro, even though it can appear, is not licensed and is thus barred. This is the case in Italian, for example, where pro is available in external argument position of tensed clauses but not in infinitives. This prediction is difficult to verify since sentences with either PRO or pro will have identical surface structures. The obvious way to verify it is to assume as we did in 1.3 that the distribution of empty categories should be the same as the distribution of lexical NPs. This then supposes that if a lexical NP cannot appear in a given position then pro, which is the null counterpart of a phonologically realized pronominal NP, cannot appear either. On the other hand in the present case of infinitival sentences the ungrammaticality of overt subjects could be attributed to the lack of nominative Case. This would not prevent pro from appearing since Case is not relevant for pro. According to this view, the contrast shown in the Italian sentences in (23) could be taken as evidence that pro might be present in Italian infinitives. However this occurrence of pro would not be licensed.

(23)a. * Mario affermava [*questa donna* non volerlo sposare].
'Mario stated [this woman not to want to marry him].'
b. *? Possiamo ritenere [*queste persone* avere sempre fatto il loro dovere].
'We can believe [these persons to have always done their duties].'
c. Affermo [di *PRO* avere sempre fatto il mio dovere].
'I state [PRO to have always done my duty].'
d. Ritengo [di *PRO* avere sempre fatto il mio dovere].
'I believe [PRO to have always done my duty].'
(Rizzi 1982)

The second possibility involves a language exhibiting inflected infinitives. In such a language, pro would be licensed in subject position of non-tensed sentences so long as there is rich subject-verb agreement. Raposo (1985;1987) reports that Portuguese and Galician fit this description.[11]

One way of verifying the availability of pro as subject of inflected infinitives is to show that inflected infinitives with empty subjects are not control structures. In other words, if the subject were PRO, we would expect it to either be coreferent with (controlled by) the matrix subject or object or to be arbitrary in reference. The sentence would therefore not be ambiguous. This is the case in English as shown by the examples in (24).

(24)a. I expect [PRO to leave].
b. I hope [PRO to arrive soon].
c. I tried [PRO to win the prize].

On the other hand, Rouveret (1980) argues that in Portuguese a construction is ambiguous between a coreferent reading and a disjoint reference reading. The only difference between the sentences in (25) and those in (24) is that the infinitive verb in Portuguese is inflected. The following examples are taken from Rouveret (1980); agreement markers are underlined; cf. footnote 11.

(25)a. Acreditam [ter*em* gastado esse dinheiro para nada].
'They think [to have spent this money for nothing].'
b. Confessam [dever*em* lhe a vida].
'They admit [to owe him life].'
c. Lamentam [não saber*em* ler].
'They regret not to know (how) to read.'
d. Acreditam [que pro têm gastado esse dinheiro para nada].
'They think [that they have spent this money for nothing].'
e. Ele diz [pro se*mos* pobres].
'He says (we) to be poor.'
f. Afirma [pro ter*em* lhe roubado este livro].
'He affirms (they) to have stolen this book from him.'
g. Dir-se-ia [pro não poder*em* carlar-se].
'One would say (they) not to be able to shut up.'

Sentence (25a) is as ambiguous as its counterpart with a tensed sentential complement in (25d). Note that the sentential complement in (25d) has no overt subject as in (25a).

If it is true that pro can be the subject of inflected infinitives, then there should be no restriction on its content. It should be possible to have a different marker on the matrix verb and on the embedded verb. This is shown in (25e-g) from Rouveret (1980).

I thus conclude that pro can in fact occupy the external argument position of inflected infinitives at least in Portuguese.

It should be clear from the preceding discussion that there must be other constructions in which pro can be licensed. The ones I reviewed here are the most well-documented ones in the generative literature, which does not constitute a guarantee that the set is exhaustive. There are languages, for instance, in which only (some) expletive (or pleonastic) pronouns can be phonologically empty. German, Irish, Malagasy and Yiddish are languages of this sort. Travis (1984, 226ff) offers an analysis of this phenomenon that is compatible with the one presented here even though, for her, pro must be properly governed, which is not the case here. Travis' account of empty expletives relies on a particular feature identification by INFL of expletive pro. I will not review this analysis here; cf. Safir (1985) for a different approach that is incompatible with my assumptions.

2.2.2. *On Agreement*

I will now concentrate on the notion "agreement" which has been repeatedly appealed to up to now. I will attempt to demonstrate that the intrinsic nature of subject clitic pronouns and their syntactic properties make them look very much like agreement markers with respect to the null subject property. This discussion will revive the much discussed problem of determining the differences and similarities between agreement markers and clitics. This problem is addressed in greater detail in Chapter 4.

In the case of languages like Italian and Spanish, it is generally assumed, as we have seen, that the verbal morphology carries agreement features that are lexically realized in such a way that this morphology is generally different for each person and number. This characteristic is often called "rich" inflection. The intuition is therefore that the agreement morphology carried by the verb stem is sufficient to recover the information that is unavailable if a lexical subject in [NP,IP] is absent. One question arises with regard to how the concept of rich inflection can be defined. This problem is especially relevant for learnability considerations. In other words, one can ask, How will the language-learner establish that its grammar includes an agreement process between the verb and an (external) argument that is "sufficient" for the licensing of pro in this argument position? In order to answer this question, one must look at the verbal paradigms of some languages. Let us compare French and Italian to show that rich inflection must be defined not only in terms of paradigm but also in terms of syntagm. That is, the agreement morphology must be looked at not only vertically – across person and number – but also horizontally – across tense and mood. Approached from such an angle the concept of rich agreement should become transparent and even trivial.

It is possible to study certain verbal paradigms in French in such a way that one would conclude that French has rich subject-verb agreement. Consider the examples in (26).

(26)	infinitive:	aller 'to go'		être 'to be'		avoir 'to have'	
	1S	vais	[vɛ]	suis	[sɥi]	ai	[e]
	2S	vas	[va]	es	[ɛ]	as	[a]
	3S	va	[va]	est	[ɛ]	a	[a]
	1P	allons	[alõ]	sommes	[som]	avons	[avõ]
	2P	allez	[ale]	êtes	[ɛt]	avez	[ave]
	3P	vont	[võ]	sont	[sõ]	ont	[õ]

Except for 2s and 3s, each of the verbs in (26) has a different ending for each person and number. That is the generally accepted definition of rich inflection. Nevertheless, this conclusion does not hold if one compares more paradigms syntagmatically as (27).

(27) infinitive: parler 'to speak'

	present		*future*		*past*	
1s	parle	[parl]	parlerai	[parlre]	parlais	[parlɛ]
2s	parles	[parl]	parleras	[parlra]	parlais	[parlɛ]
3s	parle	[parl]	parlera	[parlra]	parlait	[parlɛ]
1p	parlons	[parlõ]	parlerons	[parlrõ]	parlions	[parljõ]
2p	parlez	[parle]	parlerez	[parlre]	parliez	[parlje]
3p	parlent	[parl]	parleront	[parlrõ]	parlaient	[parlɛ]

The situation becomes confused as it is often the case that the same ending represents four different person/number combinations. It thus proves useful to compare paradigms of the same verb in different tenses.

Consider next the conjugation of the Italian verb in (28).

(28) infinitive: parlare 'to speak'

present subjunctive	1s	parli
	2s	parli
	3s	parli
	1p	parliamo
	2p	parliate
	3p	parlino

This conjugation is the same for all regular verbs ending with *-are* in Italian in present subjunctive. It would then seem natural to conclude that subject-verb agreement in Italian is not rich enough, which is the wrong conclusion. This becomes evident when we compare different paradigms of the same verb as in (29).

(29) infinitive: parlare 'to speak'

	present	*future*	*imperfect*
1s	parlo	parlerò	parlavo
2s	parli	parlerai	parlavi
3s	parla	parlerà	parlava
1p	parliamo	parleremo	parlavamo
2p	parlate	parlerete	parlavate
3p	parlano	parleranno	parlavano

Each person/number combination in (29) has its own ending for the three tenses. This constitutes, indeed, a rich agreement system.

To summarize, exposure to different paradigms in different tenses should be sufficient evidence for the language learner to establish that the language has an agreement system developed enough to provide licensing of pro in the argument position involved in the agreement process with the verb. I conclude that the problem of defining a concept such as "rich" inflection is a spurious one.

The language-learner probably determines whether the language has rich agreement at the time the various conjugations for regular verbs are established. Irregular verbs are presumably not involved in the making of such a "decision." The Recoverability Principle motivates the discovery procedure involved in determining the richness of the agreement since a simple tensed sentence with no phonologically realized subject can constitute evidence that pro is involved and that it must be licensed.

Having determined that French does not have a sufficiently developed subject-verb agreement system, it should be obvious that licensing of pro in this language – if there is such licensing – will not be done through agreement morphology on the verb. We can now return to the generalization in (6) above and ask why languages with subject clitics are never among the so-called pro-drop languages. Let us assume that subject clitics in these languages are the morphological realizations on the verb of the features necessary for the licensing of pro in subject position. Again, given the generalization in (6), it must be the case that clitics are analyzed as solely responsible for the licensing of pro and that, for some reason, they take priority over subject-verb agreement. This view makes two predictions. First, the subject clitics available in the grammar of these languages should always be obligatory if the external argument position is empty. This prediction seems to be verified and is expressed by generalization (6). The paradigms in (30)-(32) are all barred in the various languages when no subject clitic is present.

(30) Trentino[12]

2S	te + parli	*parli
3S	el/la + parla	*parla
3P	i/le + parla	*parla

(31) Fiorentino

1S	e' + vengo	*vengo
2S	tu + vieni	*vieni
3S	e'/la + viene	*viene

1P	si + viene	*viene
2P	vu + venite	*venite
3P	e'/le + vengono	*vengono

(32) French

1S	je + parle	*parle
2S	tu + parles	*parles
3S	il/elle + parle	*parle
1P	nous + parlons	*parlons
2P	vous + parlez	*parlez
3P	ils/elles + parlent	*parlent

This is so because licensing of pro subject by feature identification is done through the subject clitic. If the subject position is phonetically empty, then a clitic must be present in order to license pro. If the clitic is not present then pro is not licensed and the string is ruled out.

The second prediction concerns the coexistence of a rich system of agreement with subject clitics. If licensing of pro is done by the clitic, then the available clitics should always be present even though there is rich subject-verb agreement. I do not have access to a study comparing different paradigms in the N.It. dialects and the available data in Brandi and Cordin (1981), some of which is given in (30) and (31) above, indicates that the agreement systems in Trentino and Fiorentino seem to be quite developed. Nevertheless, as shown in (30) and (31), the clitics are always obligatory. This can be accounted for in the same way as the first prediction: the clitics must be present in order to identify pro in argument position.

Subject clitics are affixed to the verbs at least on the surface and this is expressed by the notational device (+) in the paradigms above. Notice that in (32) this affixation of clitics has the effect of disambiguating the paradigm found in (27). It can thus be said that the analysis presented here follows Taraldsen's intuition with respect to agreement markers in the sense that clitics help to recover the feature content of the missing arguments in the same way agreement markers do; cf. 1.1.1.

Let us now approach the same facts from a different angle by asking the following question: What, in (30)-(32), is the element with which the verb agrees? It could be either the clitic pronoun or the empty NP. In English, where subject pronouns are not cliticized to the verb, it seems reasonable to assume that the verb agrees with the pronoun, which is in fact just a regular nominal element in [NP,IP]. In languages with subject clitics, it also seems natural to propose that the verb agrees with the NP in external argument position. This position is empty in a construction with a subject clitic. This suggestion has the advantage of maintaining only one process of subject-verb agreement in the grammar. I will not attempt to develop the detail of this process here; I simply assume that a verb agrees with its external argument and that the subject clitic is not an external argument as proposed in Safir (1985).

We have already seen that pro can occupy the external argument position in languages with subject clitics if such a clitic licenses pro. The licensing process developed in 2.2.1 requires the licenser to govern the position occupied by the licensee pro. A subject clitic must therefore be able to govern the [NP,IP] position. The clitic nature itself of the pronouns in (30)-(32) makes it possible to assume that they are base-generated inside INFL. Cliticization operates through a late rule of Verb Raising; cf. Chapter 4 for the details concerning subject cliticization.

The structure in (33) results from the preceding discussion.

(33)

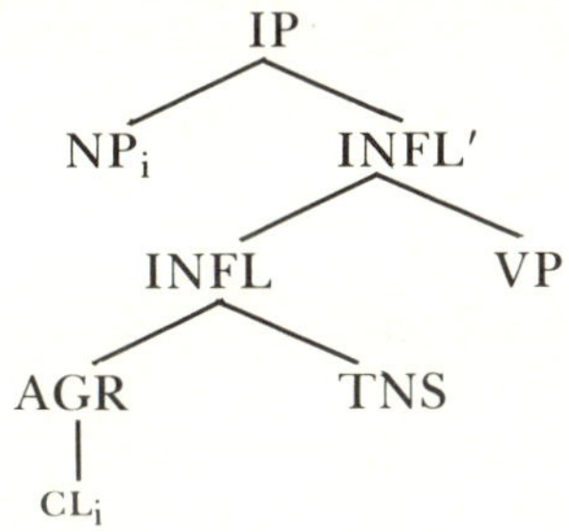

Many authors have proposed a similar structure; cf. Bouchard (1982), Brandi and Cordin (1981), Jaeggli (1984) and Rizzi (1986a) for the N.It. dialects, Everett (1985) for Pirahã, and Jaeggli (1982) and Sportiche (1983) for St.Fr.

Two possibilities arise in the case of a structure such as the one in (33): the NP position is filled with some lexical material or it is not. In the latter case, pro occupies the subject position and is licensed by the clitic in INFL.

The clitic is generated under AGR for two reasons. First, this accounts for the functional similarity between agreement markers and clitic pronouns with respect to the licensing of pro. Secondly, the TNS element serves another purpose in the grammar and may be moved into VP prior to S-structure in the languages having free inversion; cf. 2.4. This node is therefore not a good position for *Cl* to occupy in INFL.

All this has a number of implications for a theory of learnability. Since there is a primary division among languages concerning the availability of pro, the child has to decide whether or not the language has the means to license pro. This decision can be made by observing whether subject pronouns are clitics or not. Since Kayne (1972;1975), we know that there are simple tests to recognize the clitic status of pronouns. One of them is the difference in the behavior of lexical NPs and pronouns with respect to inversion in French. The contrasts between (34a) and (34b), and between (35a) and (36b) constitute positive evidence that pronouns are not in the same syntactic position as lexical NP subjects.

(34)a. Viendra -t-il?
'Will he come?'
b. Quand viendra(*-t-) Yves?
'When will Yves come?'

(35)a. Quand a-t-il dit qu'il irait?
'When did he say that he would go?'
b. *Quand a Jean dit qu'il irait?
'When did Jean say that he would go?'

(36)a. *Quand a dit-il qu'il irait?
b. Quand a dit Jean qu'il irait?

In other words, if the pronouns are never missing and are clitics, then pro can be licensed as in French, Trentino, and Fiorentino. If the pronouns are never missing and are not clitics, then pro is not available, as in English. If the pronouns are missing, then they are not clitics – according to (6) – and the presence of pro is made possible by licensing through rich verbal inflection as in Italian, Portuguese, and Spanish. The fourth possibility – pronouns are missing and are clitics – is excluded by (6).

2.2.3. *Subject Clitics and Object Clitics*

In the Government and Binding framework, the relevant aspects of which are presented in 1.1, certain differences and similarities in the behavior of elements in the sentence with respect to the principles of UG can be characterized in terms of structural properties. For example, subjects (external arguments) and objects (internal arguments) can behave differently with respect to extraction possibilities and this follows on some accounts from their relationship with the predicate. In this section, I present one major general subject/object symmetry in the relationship between clitics and the argument position they are related to. In the preceding section, it was argued that the structure in (33), where subject clitics are generated in AGR of INFL, is best suited to account for the cross-linguistic generalization in (6). More specifically, the clitic in (33) can govern the external argument position which can only be filled by pro. In turn, the clitic licenses the presence of pro. This explains why a clitic must appear in languages with subject clitics when the subject position is empty in D-structure in a tensed clause.

This account can now be extended to cover the cases of empty internal argument position in languages with object clitics. Consider the following examples from Italian, Spanish, Arabic, and French (Standard, Colloquial, and Pied Noir).[13]

(37) Italian
  a. Conosco Gianni molto bene.
     know-1s very well
     'I know Gianni very well.'
  b. *Conosco *e* molto bene.
  c. Lo_i conosco e_i molto bene.
     him
     'I know him very well.'

(38) Spanish (Jaeggli 1982)
  a. Vimos la casa de Mafalda.
     saw-1P the house of
     'We saw Mafalda's house.'
  b. *Vimos *e*.
  c. La_i vimos e_i.
     her/it
     'We saw it.'

(39) Arabic
  a. šif-it Xalid
     saw-1s
     'I saw Xalid.'
  b. *šif-it *e*.
  c. šif-t $-a_i$ e_i.
     saw-1s-him
     'I saw him.'

(40) French

a. Pierre a vidé la bouteille.
has emptied the bottle
'Pierre emptied the bottle.'

b. *Pierre a vidé *e*

c. Pierre l_i' a vidée e_i.
her/it
'Pierre emptied it.'

In these languages with object clitics, a transitive verb cannot have an internal argument represented by a base generated empty category unless an object clitic is present. This generalization patterns with the one in (6) concerning external arguments. The same explanation holds: the presence of the empty category (pro) in object position is forced by the Projection Principle. The Recoverability Principle forces the presence of a clitic in order to license the empty pronominal pro in argument position.

As far as the structure is concerned, the relationship between an object clitic and the argument associated with it must be similar to the one in (7), i.e. the clitic affixed to the verb must govern the object position. This is illustrated in (41).

(41)a.

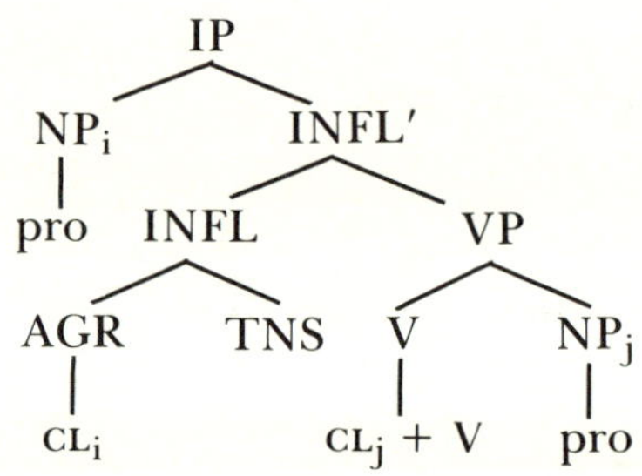

b.

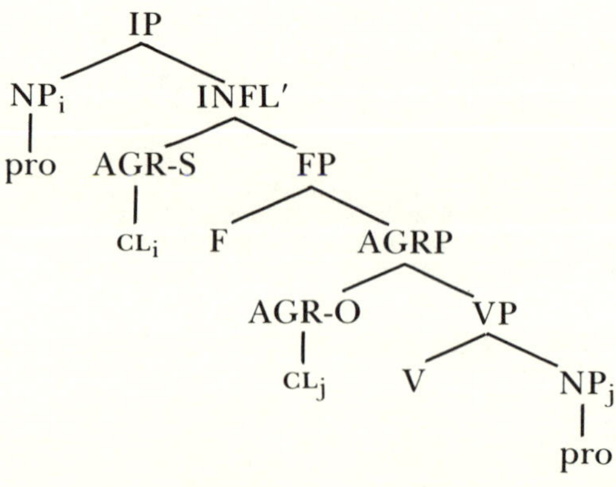

Note that this structure now seems to be generally accepted in the generative literature;[14] cf. Chomsky (1982), Zubizarreta (1982b), Sportiche (1983), and Jaeggli (1984;1986) among others. The more symmetrical structure in (41b) can be proposed based on recent work by Pollock (1988) and Chomsky (1989). Although certain aspects of the argumentation presented here would have to be modified, the main predictions remain the same, so I will continue to use the structure in (41a) in the rest of this book.

More symmetries and asymmetries, especially with respect to clitic doubling, will be discussed in chapter 3. What is important to note here is that object clitics and subject clitics can be assumed to be two distinct realizations of the same element, the syntactic function of which is to license pro in argument position. This discussion can be extended to other constructions involving clitics that have been presented in 2.2.1, namely, object of a preposition in (12) and (13), complement of a noun in (14) and (15).

2.2.4. *Base Generation of Clitics*

In the preceding discussion I have assumed that clitics in general are elements that are base generated on some head in the construction they belong to. There is some dispute in the literature as to whether clitics are base generated in the position they occupy on the surface or are moved from the argument position to which they are related. The latter approach is developed in Kayne (1975), Quicoli (1980) and Rouveret and Vergnaud (1978) whereas discussions of the former can be found mainly in Strozer (1976), Rivas (1977), Jaeggli (1982;1986), Borer (1984) and Safir (1985). I will not attempt here to review the arguments put forward to defend these two different approaches. The reader is referred, in particular, to Jaeggli (1982, 15ff) for such a review. These studies are mostly concerned with object clitics and causative constructions; I return to the latter in 2.5. In the present section, I concentrate on subject clitics and on two different theories concerning their generation across languages. I have already discussed the major difference between the two theories. In the theory, defended in Kayne (1984) and Rizzi (1986a), the differences among languages exhibiting subject clitics arise because the notion "subject clitic" is not a homogeneous one in theories of UG: what is often referred to as "cliticization" can operate in various ways across languages and even within a single particular grammar. In the theory I argue for here, the concept "subject clitic" is not grammatically spurious but refers to a very precise syntactic notion. It follows from this view that subject clitics should be base generated uniformly across languages. Cross-linguistic variation would then follow from differences in the intrinsic properties of the subject clitics in each language.[15]

Let us first review the analyses in Kayne (1984) and Rizzi (1986a), introducing the relevant data at the same time.

Kayne assumes the existence of a rule of Clitic Placement for object clitics that leaves a trace in argument position counting as an anaphor, and proposes that in the St.Fr. sentence given in (42) the subject pronoun cliticizes to the verb by a rule of cliticization which takes place in the phonological component of the grammar (PF).

(42) On a sonné.
'One has rung.'

Cliticization of object clitics, on the other hand, operates in syntax in the mapping from D-structure to S-structure. Kayne points out various facts supporting this approach. First, if *ils* in (43) is "less thoroughly" a clitic than *les*, then it is possible to account for the contrast in (44).

(43) Ils les voient.
they them see
'They see them.'

(44)a. ? Qui ça, ils?
who that they?'
'Who is that, they?'
b. *? Qui ça, les?

Secondly, there are well-known ordering constraints on clitic strings in French; cf. Kayne (1975). One of them is the requirement that subject clitics appear to the left of object clitics as the first member of the clitic string as shown in (45).

(45)a. Il le lui a dit.
he it to-him/her has told
'He has told it to him/her.'
b. *Le il lui a dit.
c. *Le lui il a dit.

If *il* cliticizes to the verb at PF while the object clitics do so prior to S-structure then the relative ordering can be accounted for.

Finally, the negative marker *ne* always appears between the inflected verb and the subject, be it a full lexical NP or a subject clitic, as shown in (46) and (47).

(46)a. Pierre n' est pas à Paris.
        neg is not at
    'Pierre is not in Paris.'
  b. Tout n'est pas à Paris.
    'Everything is not in Paris.'
  c. Il n'est pas à Paris.
    'He is not in Paris.'

(47)a. *Ne Pierre est pas à Paris.
  b. *Ne tout est pas à Paris.
  c. *Ne il est pas à Paris.

Nevertheless, in Kayne's approach the level at which the cliticization process of subject pronouns in St.Fr. applies is not unified. To illustrate, consider a sentence like the one in (48) involving Complex Inversion.[16]

(48) Cela est- il faux?
  that is it false
  'Is that false?'

In Kayne (1984) this sentence is derived as follows: the subject NP *cela* is moved to the left outside of S (=IP), the INFL node also moves out of S, the expletive pronoun is then inserted in subject position and cliticizes to INFL, crucially, at S-structure in order to properly govern the empty subject position.[17]

(49) – D-structure:
  $[_S\ [_{NP}\ \text{cela}]\ \text{INFL}\ [_{VP}\ \text{être faux}]]$
  – Verb movement to INFL:
  $[_S\ [_{NP}\ \text{cela}]\ [_{INFL}\ \text{est}]\ [_{VP}\ e\ \text{faux}]]$
  – NP movement out of S:
  $[_{NP}\ \text{cela}_i]\ [_S\ [_{NP}\ e_i]\ [_{INFL}\ \text{est}]\ [_{VP}\ e\ \text{faux}]]$
  – INFL movement out of S:
  $[_{NP}\text{cela}_i]\ [_{INFL}\text{est}_j]\ [_S\ [_{NP}e_i]\ [_{INFL}e_j]\ [_{VP}\ e\ \text{faux}]]$
  – *il* insertion:
  $[_{NP}\text{cela}_i]\ [_{INFL}\text{est}_j]\ [_S\ [_{NP}\text{il}_i]\ [_{INFL}e_j]\ [_{VP}\ e\ \text{faux}]]$
  – Leftward cliticization (= S-structure):
  $[_{NP}\text{cela}_i]\ [_{INFL}\text{est}_j\text{il}_i]\ [_S\ [_{NP}e_i]\ [_{INFL}e_j]\ [_{VP}\ e\ \text{faux}]]$

There are thus two operations of subject cliticization in French: one applying to the right in PF, the other to the left at S-structure.

Kayne (1984, 221, 230) acknowledges that it is conceivable that certain Italian dialects exhibit subject clitics that are base-generated in INFL, as suggested in Chomsky (1982) and Brandi and Cordin (1981). The reasoning behind this is that because the empty category in subject position in Italian is pro, there is no occurrence of an improperly bound empty category. The question here is why can such an analysis not be extended to all subject clitics cross-linguistically. This would have the advantage of reducing the process of subject cliticization to a single rule. Such an account appears to provide the more restrictive theory of (subject) clitics.

I want to argue here that the properties of subject clitics in St.Fr., accounted for by rightward phonological cliticization in Kayne's analysis, can be as easily handled by a "clitic in INFL" type of analysis along the lines of the one presented in (33) above. (I return to Complex Inversion in 2.4.)

Recall that we adopted a D-structure representation as in (50).

(50)

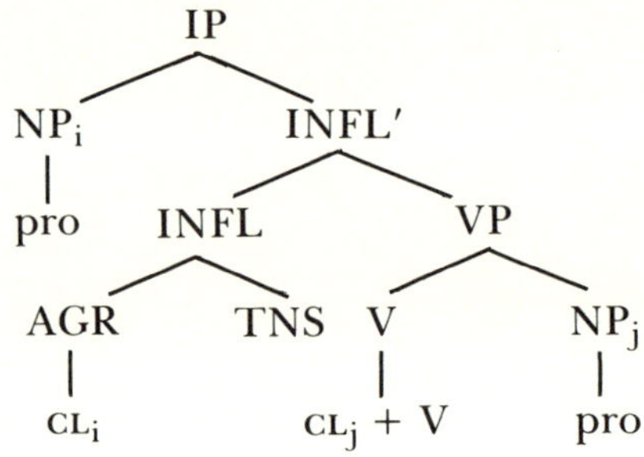

What is important to notice here is that the string $[_V Cl_j + V]$ constitutes a separate entity according to this structure.

Consider now the sentences in (43) and (44). Assuming the structure in (50), the subject clitic in INFL still constitutes a somewhat independent morpheme, at least until cliticization takes place in PF. It is thus natural to expect the relative acceptability of (44a) as opposed to (44b) since at no point in the derivation is an object clitic morphologically independent of the verb.

Consider next the contrasts in (45) regarding the position of the subject clitic with respect to object clitics. According to the structure in (50) and the analysis of cliticization, a subject clitic simply cannot be either between an object clitic and an inflected verb (45c) or between two object clitics (45b), which clearly is a desirable prediction. The only other plausible position in which a subject clitic could appear would be after $[_V CL_j + V]$. I return to this possibility in 2.4.

I now turn to the sentences in (46) and (47). Let us assume for the moment the following statement from Rizzi (1986a, 398): "the negative clitic and the subject clitic are members of the same clitic cluster, both being constituents of INFL".[18] If this is true then there is no reason to expect a difference in the relative ordering of subject clitics and *ne*, on one hand, and lexical NP subject and *ne*, on the other hand, since when a clitic is absent as in (46a,b) the negative marker would still follow the subject. Rizzi points out that this analysis of the negative marker would lead one to expect variation among languages with respect to the particular ordering of clitics and negation. This prediction is verified in various N.It. dialects.

I thus conclude that, given a structure like the one in (50), the properties of French clitics presented above can be accounted for in a natural way. There is then no reason to believe that subject clitics in French are not base generated in INFL.

The second theory to be presented here is suggested in Rizzi (1986a). Rizzi proposes that the notion "subject clitic" can refer to two very different syntactic elements. Arguments in favor of this position come from a comparison between St.Fr. and some N.It. dialects.

The crucial data involved in Rizzi's presentation concern cases of subject doubling (cf. (1) above) in these Italian dialects. A Trentino example is given in (51).

(51)a. El Gianni el magna.
DET he eat-3s
'Gianni eats.'
b. El magna.
c. *Magna.

According to Rizzi, these dialects represent an apparent intermediate case between clear examples of non null subject languages like English and clear cases of null subject languages like Standard Italian and Spanish. He provides convincing evidence for the structural representation in (52) and against the ones in (53) and (54).[19] I will not review his arguments here.

(52)
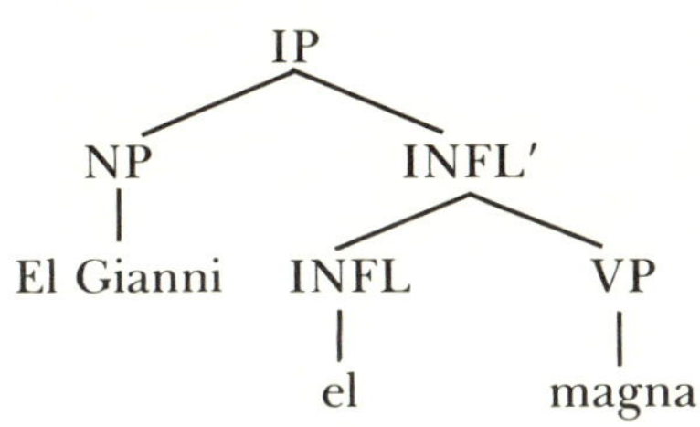

(53)

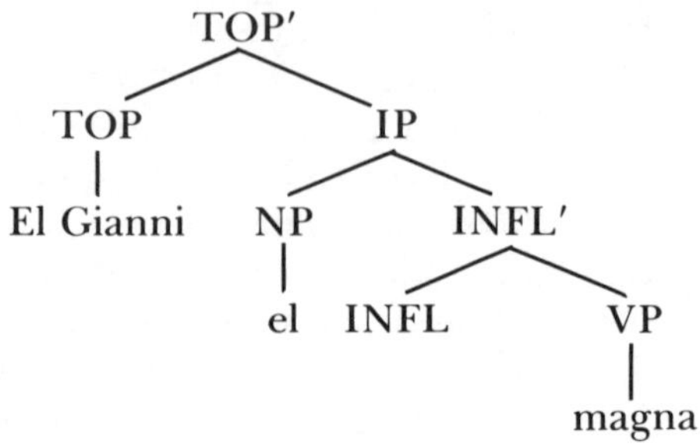

(54)

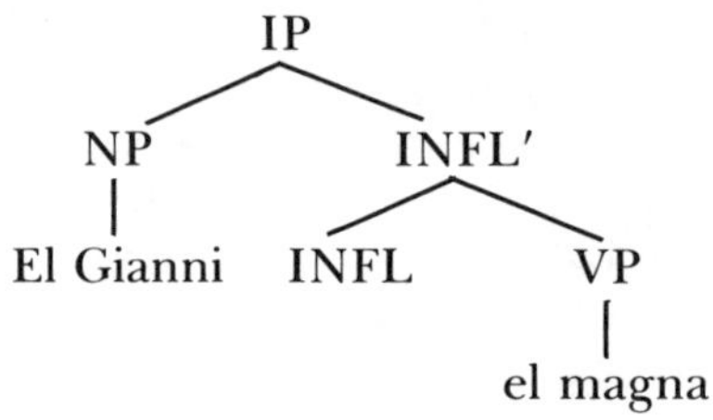

In (52) when the subject position is empty, pro is found in this position, where it is licensed – in the sense adopted here – by the rich AGR which is phonetically realized both by the subject clitic and by the verbal morphology.

What is more important for our discussion is the assumption that the two sentences in (55) have different syntactic representations in Rizzi's approach.

(55)a. Standard French
Il mange.
'He eats.'
b. Trentino
El magna.
'He eats.'

Let us review the reasoning behind this. First, there is no subject doubling in Standard French, compare (56a) to (51a).

(56) Standard French
a. *Jean il mange.
b. Jean mange.
c. Il mange.

Unless there is a pause between *Jean* and *il*, i.e. unless left-dislocation has occurred, (56a) is ungrammatical. Furthermore, even for speakers who find (56a) acceptable without a pause, this type of construction becomes ungrammatical with a quantifier as subject. This is not the case in Trentino.

(57)a. Standard French

*Personne	il	ne	mange.
nobody	he	NEG	eat-3s

'Nobody eats.'

b. Trentino

Qualche	putel	l'	é	vegnu.
some	boys	they	have	come

'Some boys came.'

In Rizzi's approach, this would mean that a structure like the one in (53) would be best suited for St.Fr.; cf. also Brandi and Cordin (1981) and Kayne (1984) for similar views.

Secondly, some of the N.It. dialects have gaps in their paradigm of subject clitics, but this is not the case in St.Fr., as shown by the paradigms in (58).

(58)	Trentino		Standard French		Fiorentino	
1S	ø	vegno	je	viens	e'	vengo
2S	te	vegni	tu	viens	tu	vieni
3SM	el	ven	il	vient	e'	viene
3SF	la	ven	elle	vient	la	viene
1P	ø	vegnim	nous	venons	si	viene
2P	ø	vegní	vous	venez	vu'	venite
3PM	i	ven	ils	viennent	e'	vengono
3PF	le	ven	elles	viennent	le	vengono

Other N.It. dialects do not have such gaps. Compare the Fiorentino paradigm in (58c) with the Trentino one in (58a) and the St.Fr. one in (58b).

Rizzi (1986a, 402) proposes to account for the difference between Fiorentino and Trentino by assuming a parameter involving the possibility that spelling out AGR in INFL is "obligatory or optional, or impossible for different specifications of person and number in the [N. It.–Y.R.] dialects, and always impossible in Standard Italian." Rizzi notes that this explanation does not hold in St.Fr.. The structure in (53) would explain why Rizzi's account cannot apply since he assumes subject clitics are in [NP,IP] in St.Fr.. Therefore, the paradigms in (58b,c) are explained in two quite different ways even though they seem identical. I believe this to be an arbitrary distinction and an undesirable consequence of Rizzi's approach.

Third, consider the contrasts in (59) and (60).

(59) Standard French
    a. Elle chante et danse.
       'She sings and dances.'
    b.?? Elle chante et elle danse.

(60) Trentino
    a. *La canta e balla.
       'She sings and dances.'
    b. La canta e la balla.

Rizzi suggests that a zero pro-form is not allowed in AGR under coordination. Therefore, if the structure in (52) is assumed for Trentino, then (60a) is excluded. On the other hand, if (53) is the correct representation for French, then (59a) will be grammatical since there is no zero pro-form in AGR of the second conjunct. The oddity of (59b) remains unaccounted for.

Finally, the type of inversion exemplified in (61) for St. Fr. obeys two constraints: the verb must be unaccusative or passivized and the inverted subject must be indefinite; cf. Safir (1985) for a discussion.

(61)a. Il est arrivé une/*la fille.
       there is arrived a/the girl
       'There arrived a girl.'
    b. Il a été arrêté un/*l' étudiant.
       there has been arrested a/the student
       'There was arrested a student.'

A similar type of construction in the N.It. dialects does not involve these constraints.

(62) Fiorentino
    Gl' é venuto le ragazze.
    there has come the girls
    'There came the girls.'

According to Rizzi, this contrast is to be related to the fact that French is not a null subject language and that inversion is more restricted in this language. In short, the postverbal NP cannot receive nominative Case from INFL since it is the expletive clitic *il* which gets it.

The only other way for the NP to be Case-marked is if it is the object of an unaccusative or passivized verb which assigns partitive Case. Partitive Case, by nature, requires indefiniteness; cf. Belletti (1985). In the Italian dialects, the subject position is empty making it possible for the postverbal NP to receive nominative Case. There is thus no constraint on the nature of the verb or on the type of NP subject that can be involved in inversion constructions.

In summary, in Rizzi's approach, subject clitics in St.Fr. are base generated in [NP,IP] position whereas in the N. It. dialects they appear in INFL in D-structure. The N.It. dialects are null subject languages but St.Fr. is not. The four differences above between St.Fr. and the Italian dialects follow from these assumptions, i.e. from structural reasons.

I want to argue here that Rizzi's conclusions should be revised given the generalization in (6), the discussion in 2.2.3, and some data from two French dialects that will be introduced below. In other words, it would seem preferable to assume the structure in (52) – equivalent to (33), i.e. subject clitics in INFL – for all the languages with subject clitics. I will discuss the coordination facts in greater detail in Chapter 4 and the inversion facts in 2.4.

Colloquial French and Pied Noir French exhibit optional subject doubling. The sentences in (63) are grammatical in both dialects.

(63)a. Jean/l'homme vient.
'Jean/the man comes.'
b. Il vient.
'He comes.'
c. Jean/l'homme il vient.

The only difference between these two dialects with regard to subject doubling is the existence in C.Fr. but not in PN Fr. of a constraint on the nature of the doubled NP. In C.Fr., the NP cannot be indefinite nor can it be quantified. This does not hold for PN Fr.

(64) Colloquial French

a.	*Un	homme	il	vient.
	a	man	he	comes
b.	*Chaque	femme	elle	vient.
	each	woman	she	comes

(65) Pied Noir French
a. Un homme il vient.
b. Chaque femme elle vient.

I defer detailed analysis of this construction to Chapter 3. The reader is referred to 3.1.3.1 for evidence that subject doubling is indeed available in C.Fr. If it is true that the N. It. dialects represent, in some sense, an intermediate case between fully null subject languages (St. It.) and languages like St. Fr. then it is reasonable to assume that C.Fr. and PN Fr. are intermediate cases between St. Fr. and the N.It. dialects. To illustrate this, let us classify the different types of dialects or languages that are referred to here. The term "restricted subject doubling" refers to the constraint on the doubled NP shown in (64).

(66)a. Standard French
+ subject clitics
− subject doubling
b. Colloquial French
+ subject clitics
+ optional subject doubling (restricted)
c. Pied Noir French
+ subject clitics
+ optional subject doubling (non-restricted)
d. Fiorentino
+ subject clitics
+ obligatory subject doubling (non-restricted)

What I want to suggest here is that under an account in terms of structural differences with respect to where the subject clitic is generated, the cross-linguistic variation expressed in (66) would look accidental and arbitrary. One would have to assume that C.Fr. and St.Fr. have their subject clitics in [NP,IP] whereas PN Fr. and Fiorentino have the same clitics in INFL. This could not be otherwise given Rizzi's discussion of (57) above, and he is forced to say that C.Fr. has the structure in (53) for subject doubling constructions.

But this causes problems because one would further have to explain why subject doubling is possible in C.Fr. but not in St.Fr., on the one hand, and why subject doubling is obligatory in Fiorentino and optional in PN Fr., on the other hand.

If one decides to classify C.Fr., PN Fr. and Fiorentino as languages with subject clitics in INFL but St.Fr. as one where subject clitics are in [NP,IP], one would run into the difficulty of explaining the optionality of subject doubling and its restricted character in C.Fr. This might also cause learnability problems since it is not obvious how the child could distinguish between the two groups once the presence of subject clitics has been detected.

On the other hand, an analysis in which the languages and dialects in (66) are all analyzed as having the same structural representation, i.e. where subject clitics are base-generated in INFL, makes the dialectal variation more natural given differences in the properties of the clitics involved with respect to other independently motivated principles of UG. In other words, this type of analysis considers the term "subject clitic" to refer to the same specific grammatical element in all the languages exhibiting it.

Now, with respect to the generalization in (6) and the discussion in 2.2.3, the analysis just outlined would be able to relate various properties of constructions with subject clitics to similar properties of constructions with object clitics. It is a well-known fact that some languages with object clitics allow object doubling. Interestingly, the distribution of this property is equivalent in many respects to the one in (66), as illustrated in (67). I do not have any examples of languages requiring obligatory object doubling.

(67)a. Standard/Colloquial French, Italian
+ object clitics
− object doubling
b. River Plate Spanish A, Pied Noir French[20]
+ object clitics
+ optional object doubling (restricted)
c. River Plate Spanish B[21]
+ object clitics
+ optional object doubling (non-restricted)

It would thus seem to make sense to try to relate object clitics and subject clitics and, in fact, assume that they are two different realizations of the same grammatical element having the function of licensing pro in argument position.

Finally, the unified analysis of subject clitics proposed here does not require that (58b) and (58c) be analyzed in different ways. Rizzi's (1986a) suggestion concerning the obligatoriness, optionality or impossibility of spelling-out AGR still holds. Whatever makes the Fiorentino paradigm what it is also makes possible the St. Fr. paradigm.

2.2.5. *Case and Θ-Role*

I will now detail my assumptions concerning Case and Θ-role assignment in clitic constructions. To do so, it would be ideal to maintain the same line of reasoning with respect to the subject/object symmetry developed in 2.2.3. It is thus useful to return to our basic assumptions concerning the relationship between a clitic and the argument to which it is related.

We have seen in 2.2.1 that pro can be licensed across languages in the positions where a lexical pronoun can appear. It is thus natural to continue to assume that a Θ-role is assigned to an argument position as is usually the case in the Government and Binding framework. Now, what are the conditions on Θ-theory? The Θ-criterion is one of them; a simplified definition is given in (68).

(68) Theta Criterion
Each argument bears one and only one Θ-role, and each Θ-role is assigned to one and only one argument.
(Chomsky 1981, 36)

A clitic and its related argument are coindexed for licensing purposes. For this reason, I adopt Borer's (1984) suggestion that the position occupied by the clitic has no status (A or $\bar{A}$) with respect to the Θ-criterion. If it were in an A-position, the verb would select two argument positions (SCL and NP) and assign the same Θ-role to both. This is excluded by the Θ-criterion; but see di Sciullo (1982) for whom clitics are $\bar{A}$-binders for a variable in A-position. In fact, Sportiche (1983) argues that the problem of determining an A- or $\bar{A}$-status for clitics is a vacuous one. In the approach outlined here, this remark holds since the problem amounts to asking whether or not the agreement marker *-s* in (69) is in an A- or $\bar{A}$-position.

(69) John leaves tomorrow.

Jaeggli (1986) also adopts the same type of approach.

The question that remains concerns Case-assignment. Before considering the examples of doubling, I will simply assume, for the moment, following Borer (1984) and Jaeggli (1986), that a clitic absorbs (or is a spell-out of) Case. If this is so then it must not be that pro is licensed in the domain of Case-assignment, as assumed in Rizzi (1986b), since in the example given in (70) the subject position is not assigned Case; cf. 2.2.1.

(70) $[_{IP} [_{NP}$ pro$_i] [_{I'} [_{INFL}$ il$_i] [_{VP}$ vient$]]]$

The assumption that Case is absorbed by a clitic will be qualified in chapter 3 when we discuss clitic doubling.

Finally, note that this analysis of Case and Θ-role applies for both subject clitics and object clitics in a configuration like the one in (50). Moreover, this analysis seems to be a plausible candidate for parametric variation of the type illustrated in (66) and (67) above. This possibility is explored in chapter 3 together with the exact relationship in terms of a chain holding between a clitic and a related argument position.

2.3. ON THE NON-EXISTENCE OF THE NULL SUBJECT PARAMETER

It is now time to put the results of the preceding sections in a more general perspective. In very general terms, the goal of linguistic theory, as conceived of here, is to account for variation and learnability. It is therefore desirable to try to attain a maximum of generality in the postulation of principles and parameters. A good example of this was presented in 1.2.1 where I reviewed the evolution of the studies on the null subject phenomenon. We have seen that these studies have concentrated on particular, seemingly isolated, characteristics of different languages: null subject property, *that-t* effects, inversion, etc. In Chomsky (1981), these characteristics were merged and assumed to be the result of a unique parameter. The positive setting of this parameter by the language learner, in a sense, provokes a chain of events, the result of which is that a certain number of the properties of the language follow automatically from the setting in a given way of a single parameter. The goal of linguistic theory seemed to be achieved in this particular area of research.

However, subsequent research on other languages within the new perspective developed in Chomsky (1981), has revealed that the cluster of properties associated with the null subject parameter is not as immutable as was first believed or hoped. Safir (1985), in particular, argues that given the existence of the N.It. dialects and Portuguese, the null subject property and free inversion can not be directly related to each other. Some languages appear to exhibit free inversion without null subjects and vice versa. It thus seems as if these two phenomena would be better analysed as the result of two distinct parameters.

In Chomsky (1982), with the introduction into UG of an empty category corresponding to an overt pronominal element, the null subject parameter, as formulated in Chomsky (1981), is no longer necessary. The existence of null subjects can be more closely tied to the richness of the verbal inflection indicating subject-verb agreement, as it is in Taraldsen (1978).

The result of this evolutionary process is that the cluster of properties associated with the null subject phenomenon is reduced to a minimum. No "null subject parameter" needs to be formulated.

This is a desirable result for several reasons. Most importantly, the availability of pro renders the null subject property much more deeply rooted in UG; there are consequently fewer parameters to set. The possibility of having null subjects in a given language follows automatically from the theory of empty categories, i.e. a theory of recoverability. It is very similar to the possibility of having empty subjects in non-tensed sentences in English, for example. In this particular instance of a more general theory of recoverability nothing needs to be stipulated: PRO must be ungoverned, which follows from binding theory, and the only argument position that is not governed in English, if one assumes the Extended Projection Principle, is the subject position of infinitival clauses. No parameter needs to be set.

The same is true of pro. Binding theory makes it generally available, but it appears only in governed positions and only if it is licensed. The task of the learner is to determine whether or not pro is licensed in particular stuctural contexts and this can be done through positive evidence. On one hand, verbal paradigms like the ones presented in 2.2.2 are directly available in the language a child is exposed to and, on the other hand, simple sentences, as in (71), are presumably also at the learner's disposal.

(71)a. Italian
Ho trovato il libro.
'(I) found the book.'
b. Spanish
Estamos cansadísimos.
'(We) are very tired.'
c. Portuguese
Disse que tenha lido esse livro.
'(She) says that (she) has read that book.'

Furthermore, according to studies in Bouchard (1982), Brandi and Cordin (1981), Chomsky (1982), Everett (1985), Jaeggli (1984;1986), Rizzi (1986a), Sportiche (1983), Zubizarreta (1982b), and the present book, it now turns out that (subject and object) clitics can also be proper licensers for pro. This makes interesting predictions, especially with regard to clitic doubling. So, more languages are now considered to be null subject or null object languages and variation can be found in the way in which the null argument is recovered, i.e. how pro is licensed. Thus it is now possible to establish a relationship between, say, a language with subject clitics and a language with rich subject-verb agreement or between a language with object clitics and a language with rich object-verb agreement. It now seems preferable to refer to these languages as "null argument languages," following McCloskey and Hale (1984).

The consequences of this shift in linguistic theory will undoubtedly be numerous, and more important questions will be raised. One of these questions concerns the distinction between agreement markers and clitic pronouns; part of chapter 4 is devoted to this issue. Notice also that such a perspective provokes a shift in the conception of markedness as it applies to the property under study here. One must ask not why Italian has empty arguments but why this property is not attested in languages like English as well. In other words, why does the grammar of English make no use of an empty category available in UG? The answer is probably that English does not have any licensing device for pro. More generally, the Recoverability Principle is violated in the case of this particular empty category.[22]

In the remainder of this chapter, I concentrate on related issues concerning free inversion and causatives constructions. The clitic doubling phenomenon is explored in chapter 3.

2.4. FREE INVERSION

There are various types of subject inversion in the languages under study here that need some discussion either because of the problems they pose for the analysis of subject clitics proposed here or because they raise interesting questions concerning parametric variation. In this section, I will treat only one of them, namely, free inversion. French complex inversion will be discussed in 3.4 because of its apparent similarity to clitic doubling.

It was pointed out in 1.2 that, at the onset of the studies on null subject languages in the generative literature, the possibility of having empty subjects was parametrically related to another property, namely, that of having lexical NP subjects appearing quite freely in postverbal position; cf. Jaeggli (1982) and Chomsky (1981). By freeing the external argument position from government by INFL through a rule of INFL-lowering applying prior to S-structure, Chomsky (1981) assumes that nominative Case can be assigned to a postverbal NP generated, crucially, inside VP. This operation is illustrated in (72).

(72) (→ Case-assignment)

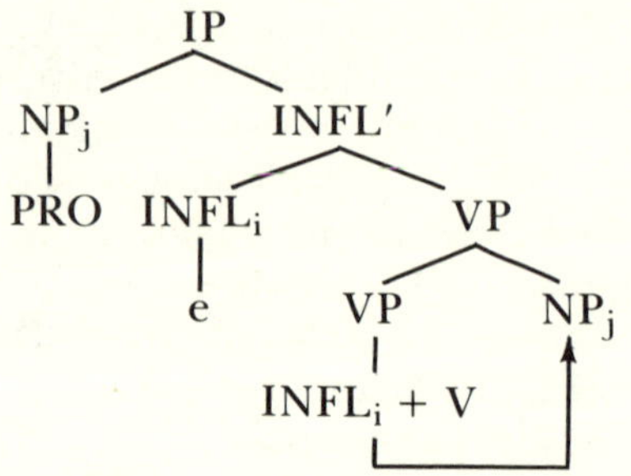

Thus, the presence of empty preverbal subjects (PRO) automatically allows for postverbal lexical NP subjects as in (73) and that is precisely how empty subjects and free inversion are to be related.[23]

(73) Italian

a. Ho trovato il libro.
'(I) found the book.'

b. Ha mangiato Giovanni.
has eaten
'Giovanni ate.'

But, as pointed out in Chao (1981), Portuguese causes problems for this type of approach. This language allows for empty subjects, exactly as in Italian, without any possibility for free inversion. Compare (74) to (73). Example (74a) is from Safir (1985) and (74b,c) are from Perlmutter (1976), cited in Safir (1985).

(74) Portuguese

a. Disse que tenha lido esse livro.
say-3s that have-3s read that book
'She says that she has read that book.'

b. Deus existe.
'God exists.'

c. *Existe Deus.

The cases of inversion found in Portuguese are closer to the French impersonal construction in that they show some kind of definiteness effect in the sense of Safir (1985) as illustrated in (75) from Perlmutter (1976).

(75)a. Existem homens capazes de matar até as aves canoras.
'There exist men capable of killing even song birds.'

b. Não existe tal ilha.
'No such island exists.'

It is sufficient to notice here the fact that Portuguese has null subjects but no free inversion and that, consequently, the two properties should be accounted for independently.[24] I adopt this line of reasoning here since it is consistent with the approach to null arguments developed so far. Since the possibility of having null subjects is allowed for by the binding theory and the Extended Projection Principle, as pointed out in 2.3, there is no obvious reason why free inversion and null subjects should be the result of a single parameter. Free inversion should therefore be the result of an independent parameter to be fixed by the language learner on the basis of the available evidence.

Now, there are many ways in which this idea can be implemented. Some of them have been proposed in the literature. My intention here is not to make an extensive review of the different options but merely to explore different alternatives.

The first option is described above and in 1.2.1 and is proposed in Chomsky (1981). We have already seen that it fails to make a clear distinction between free inversion and null subjects. According to this analysis, everything else being equal, any language allowing null subjects should also allow free inversion. However, this is an incorrect prediction.

The second option, presented in Rizzi (1982) and adopted by Picallo (1985), assumes a rule moving the subject to a postverbal position. The postverbal NP and pro in subject position are cosuperscripted and form a chain which is assigned Case. The external Θ-role is transmitted by pro to the NP; pro is therefore expletive. (→ Case-assignment)

(76)

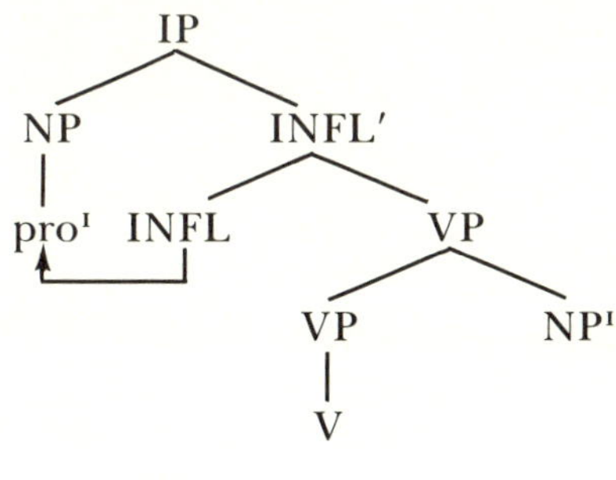

Such a movement analysis makes some predictions with respect to languages with subject clitics and free inversion. If we adopt an analysis of subject clitics that base generates them in INFL then we predict that a preverbal subject clitic would agree with the postverbal subject, and, further, that the verb would show an ending that corresponds to the person and number of the subject. This is an incorrect prediction as we will see later in this section.

The third possibility is to simply assume that an external argument can be generated through X′ theory (or alternatively by Case adjacency requirements) to the right of VP dominated by IP; cf.2.5. At least two alternative structures arise, as in (77) and (78), from Travis (1984, 190).[25]

(77)

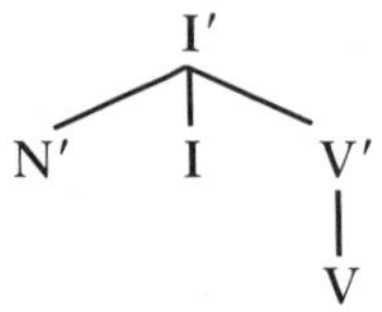

(78)

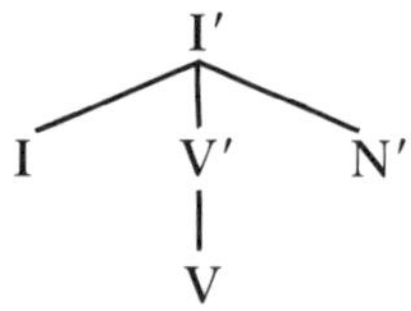

Here again languages with subject clitics and free inversion cause problems since a subject clitic in INFL in (78) would not be associated, in a sense to be made precise in 3.3, with an empty pronominal in argument position. Given the approach developed so far, it should be clear that the possibility of such an association between clitics and argument position must be excluded since if clitics can appear freely on heads without an associated argument then most generalizations concerning their appearance are lost. The second problem with this option is that Portuguese should either exhibit free inversion, which is not the case, or should not be a null subject language, which is not the case either. This is so because, as Travis (1984, 207) claims,

> the only reason that pro-drop seems to co-occur with subject-inversion is that most pro-drop languages being studied are S-I-V-O languages and S-I-V-O languages with subject-inversion do not have a properly governed subject position. Without such a position, pro-drop would leave an unidentified gap. If an S-I-V-O language has an I-V-O-S variant, however, there is the possibility of having an identified gap in subject position and features can be supplied to the gap by INFL.

Travis examines Irish which is a null subject language without free inversion but, following McCloskey and Hale (1984), she assumes it to be I-S-O-V. Irish thus does not need to appeal to free inversion in order to have an identified gap, i.e. in order to have null subjects.

Now Portuguese, contrary to Italian or Spanish, does not have an I-V-O-S variant as pointed out before (i.e. it does not have free inversion). Therefore, if null subjects are allowed, they must arise in some other manner, for example, from an I-S-O-V variant as in Irish. This variant would need to be argued for independently.

This last option is proposed in Safir (1985). Safir's basic insight, of which a variant is adopted here, is, I believe, the right one. This insight is that the free inversion possibility is the result of an independent parameter and it predicts that there exist languages with null subjects but no free inversion, like Portuguese. For arguments against the details of Safir's approach see 3.1.2.

The account of the free inversion property that I want to propose here, following an idea by M. Rochemont (personal communication), is a synthesis of the four theories above.

Let us first assume that nominative Case is assigned by the +TNS element of INFL, not by AGR as assumed in Chomsky (1981).[26] Suppose further that free inversion is tied, as in Chomsky (1981), to the possibility of having the nominative Case assigner in VP at the level where Case is assigned or checked, that is, presumably S-structure. The [+TNS] node is moved in syntax by a rule of hopping or lowering similar to Chomsky's rule R. This is a parameterized possibility: in languages where TNS may move in syntax, free inversion is possible whereas in those where it cannot move prior to S-structure, free inversion is excluded since nominative Case cannot be assigned directly to a postverbal base generated NP under VP. This is the case in Portuguese and French. According to Safir's (1985) analysis of the definiteness effect, the fact that a definite NP can appear in an inverted position can be considered to be a trigger experience sufficient to deduce that there is free inversion in a particular language.[27]

The S-structure representation of a sentence with free inversion is thus as in (79). (→ Case-assignment)

(79)

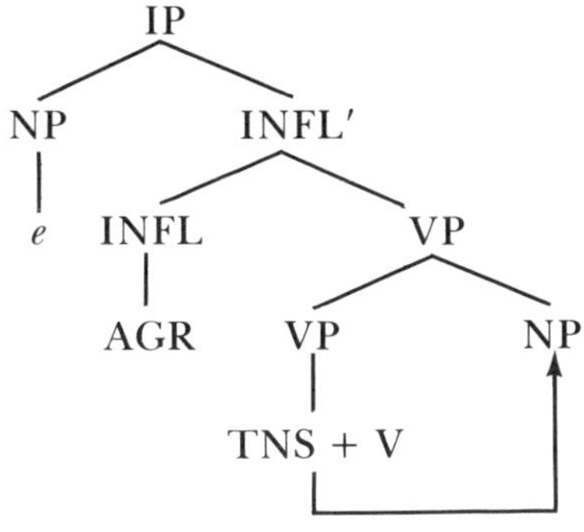

Three questions arise with regard to the representation in (79):

A. How is the external Θ-role assigned?
B. What does the verb agree with?
C. What is the status of the empty category in [NP,IP]?

In order to answer these questions, it will be necessary to compare two specific realizations of free inversion. One has already been presented – the St.It. type of free inversion in which the verb agrees, apparently, with the postverbal subject. The other is the type found in the N.It. dialects, namely, the type involving subject clitics. The major difference between the N.It. dialects and St.It. with respect to free inversion is that, in the former, the verb does not agree with the postverbal NP as shown in Brandi and Cordin (1981). In Fiorentino, for example, when free inversion occurs the neuter clitic *e* (*gl* in front of a vowel; translated as English *it*) appears in a preverbal position as in (80).[28]

(80)a. *E'* viene le ragazze.
it come-3s the girls
'The girls come.'
b. *Gl'* ha telefonato una ragazza.
it have-3s phoned a girl
'The girl phoned.'

Compare the inflection on the verb in (80) to that in (81) where the subject is in preverbal position and is accompanied by a personal clitic pronoun.

(81) Le ragazze *le* vengano.
the girls they come-3P
'The girls come."

A free inversion construction with a preverbal coreferential clitic pronoun in which the verb agrees with the postverbal subject is ungrammatical as shown in (82).

(82) **Le* vengano le ragazze.
they come-3P the girls
'The girls come.'

In Trentino there is a gap in the paradigm of subject clitics corresponding to the neuter clitic; cf. 2.2.4. We do not, therefore, find any preverbal clitic in the Trentino free inversion construction but the verb does not agree with the postverbal subject as (83) illustrates.

(83)a. 'Na putela *l'* ei vegnuda.
a girl she has come
'A girl came.'
b. *L'ei vegnuda 'na putela.
c. E vegnu 'na putela.

The same is true, in Trentino, for impersonal verbs (84), extraposition constructions (85), and impersonal passives (86).

(84) (**El*) par che el Mario *el* sia partí.
'(It) seems that Mario left.'

(85) (**El*) sará meio 'nar via.
'(It) will be better to go away.'

(86)a. E stá trová 'na borsa.
has been found a bag
'A bag was found.'
b. *L'ei stada trovada 'na borsa.

These facts suggest some answers to the questions raised previously. Let us assume the more detailed representation of (79) found in (87) for the Fiorentino sentence in (80).

(87)

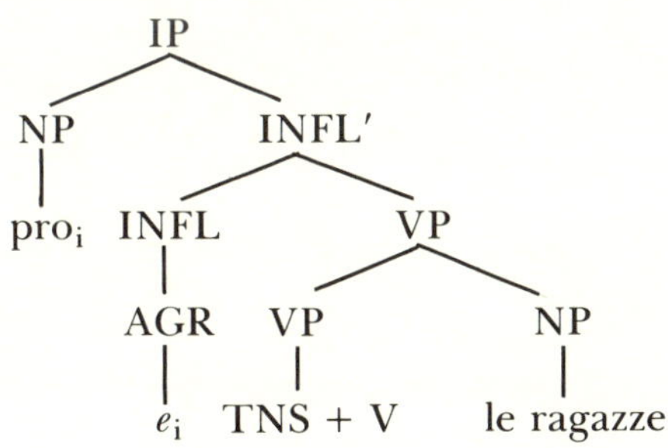

In answer to (C), a plausible candidate for the empty external argument position is pro licensed by the subject clitic *e* in INFL.

To answer question (B), we can maintain our assumption that the verbs does not agree with the inverted subject by saying that it agrees rather with the external argument position which is 3s since it is licensed by the expletive clitic in INFL. This clitic is 3s.

The process of external Θ-role assignment is less clear. The only unquestionable fact is that pro in [NP,IP] does not bear a Θ-role since it is expletive. Evidence for this comes from the Trentino data in (84)-(86) where in (84), for instance, the verb does not assign a Θ-role to the external argument position, and the behavior of the verb with respect to agreement is the same as in the free inversion construction.

There are at least two ways in which the postverbal NP in (87) can receive the external Θ-role normally assigned to the external argument position. The first is equivalent to Rizzi's (1982) proposal introduced above for St. It.: the postverbal NP receives its Θ-role from being in a chain with the preverbal pro; cf. also Belletti and Rizzi (1981, 136) for a variant of this analysis. Pro has no Θ-role and is thus expletive. This solution necessitates the formation of a "concrete" chain containing pro and the postverbal NP. This chain can be formed either by cosuperscripting or by cosubscripting the two members. This solution causes some potential problems for binding theory that are discussed extensively in Safir (1985) and which I will not review here.

The second possibility, suggested in Safir (1985, 188), is to say that the postverbal NP can somehow receive a Θ-role directly in this position. Safir's solution involves assignment of an external Θ-role to any sister of VP as long as the Θ-criterion is not violated, i.e. as long as there is only one argument for a given Θ-role. In the structures in (88) and the one in (87) both the VP-adjoined NP and the [NP,S] positions are sisters of VP and can receive the external Θ-role if any.[29]

(88)a.

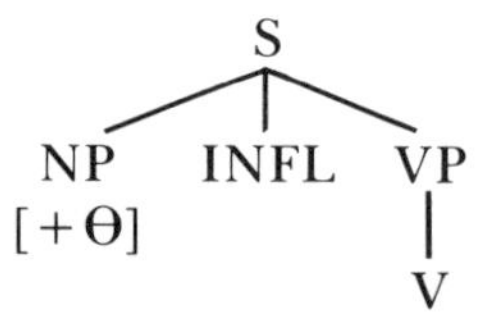

b.

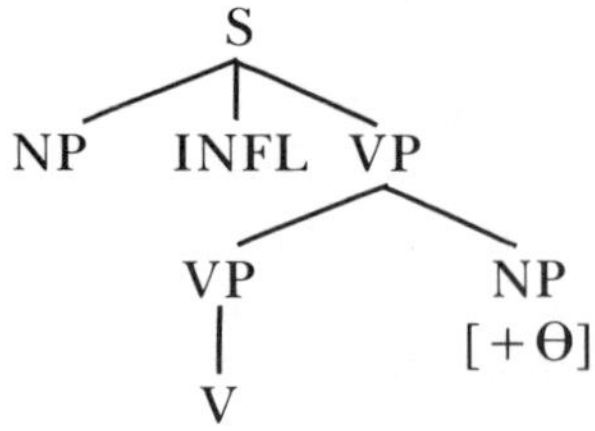

To summarize, the postverbal NP receives a Θ-role and is Case-marked directly in its position. The preverbal subject position is filled by an expletive pro licensed by an impersonal subject clitic in INFL. The verb agrees with pro in its external argument position. This type of construction is made possible in the first place by the parametrized option of moving the nominative Case assigner +TNS into VP prior to S-structure.[30]

The last issue to be discussed here concerns verbal agreement in free inversion constructions in St. It. I have already pointed out that St. It. differs from its Northern dialects in that the verb agrees with the postverbal NP.[31] Brandi and Cordin (1981) show that languages vary with respect to feature sharing between a postverbal subject and the material in preverbal position. They base this observation on the comparisons shown in the table in (89).

(89)

	+sharing property	−sharing property
Italian	Vengono le ragazze	*Viene le ragazze
English	There come some girls	*There comes some girls
French	*Il arrivent des filles	Il arrive des filles
Fiorentino	*Le vengano le ragazze	E' viene le ragazze
Trentino	*Le ven le putele	Ven le putele

(Brandi and Cordin 1981, 15)

They correctly remark that this property is independent from the null subject property since Italian and English behave in the same way. Notice that there is another correlation to be established in (89). The three languages which do not have the [+sharing] property are the ones with subject clitics; the two others have no such clitics. Data limitation together with the inconsistency regarding the types of inversion found in (89) – inversions in French and English are not "free" – prevent me from suggesting a systematic generalization. I can nevertheless speculate that the absence of a sharing property can be attributed to the presence of subject clitics in a language. Alternatively, it can be suggested that subject clitics block whatever process is involved in the feature sharing property.

Both suggestions are consistent with the proposal that subject clitics license pro in external argument position for the following reason. If it is true that free inversion is a base generated construction and that pro occupies the [NP,IP] position, then there needs to be a subject clitic in INFL in order to license it. Pro is expletive and languages with subject clitics may have one such clitic that is used for this purpose (French *il*, Fiorentino *e*, Trentino ϕ). The features of pro are determined by this clitic such that the impersonal clitic, 3s in the languages under study here, makes pro 3s as well. In languages without subject clitics (English, Italian), either the features of pro are determined by verbal agreement (Italian) or pro is not available (English); but see the discussion in the introduction with respect to English. This opens the way to a different process of agreement when a postverbal NP subject is involved in the construction, which would account for the contrasts in (89).

2.5. CAUSATIVES

Another type of construction that seems to cause problems for an analysis of object clitics in which they are base-generated on the verbal head and license pro in a Θ-marked internal argument position is illustrated by a sentence like (90b) in French, usually referred to as causative.

(90)a. Marie a fait manger la tarte à l' enfant.
has made to + eat the pie to the child
'Marie made the child eat the pie.'
b. Marie l_i' a fait manger pro_i à l'enfant.
c. *Marie a fait la_i manger pro_i à l'enfant.

The problem is that the object clitic (90b) does not enter, at first sight, into the very local relationship described in 2.2.3. This is thus a construction in which the normal pattern of cliticization in (90c) is ungrammatical.

The aim of this section is not to provide a complete analysis of Romance causatives. This construction, which has been extensively analyzed in the literature, raises interesting problems for syntactic theory. The reader is referred in particular to Aissen (1974;1979), Emonds (1978), Goodall (1984), Jaeggli (1981), Kayne (1975), Manzini (1983), Rouveret and Vergnaud (1980), and Zubizarreta (1982b).

I am concerned exclusively with the *faire* + infinitive construction where the non-tensed verb is transitive.[32] The same construction with intransitives is illustrated in (91).

(91)a. J'ai fait sortir Jean.
I have made to + go
'I made Jean go out.'
b. Ce film fait pleurer les gens.
this movie makes to + cry the people
'This movie makes people cry.'

c. Marie a fait entrer le chat.
has made to + enter the cat
'Marie made the cat come in.'

The subject of the infinitive in the sentences in (91) acts like a direct object with respect to cliticization.

(92)a. Je *l'*ai fait sortir.
b. Ce film *les* fait pleurer.
c. Marie *l'*a fait entrer.

When the verb of the infinitive is transitive, as in (93), the subject cliticizes to *faire* as an indirect object as shown by the dative clitics in (94).

(93)a. Marie a fait manger la tarte à l'enfant. (=90a)
b. Il a fait boire du vin au chat.
he has made to + drink some wine to + the cat
'He made the cat drink some wine.'
c. Il a fait signer le document à sa patronne.
he has made to + sign the document to his boss
'He made his boss sign the document.'

(94)a. Elle *lui* a fait manger la tarte.
b. Il *lui* a fait boire du vin.
c. Il *lui* a fait signer le document.

In regular constructions with an embedded tensed clause, the object of the embedded verb is cliticized onto this verb (95b), and not on any other verb (95c).

(95)a. Il exige que la patronne signe le document.
'He demands that the boss signs the document.'
b. Il exige que la patronne *le* signe.
c. *Il *l'*exige que la patronne signe.

In causative constructions, though, it is generally the case that the object cliticizes onto *faire* or its auxiliary verb, and not on the non-tensed verb as shown in (96).

(96)a. Elle *l'*a fait manger à l'enfant.
b. Elle *en* a fait boire au chat.
c. Il *l'* a fait signer à la patronne.

(97)a. ?*Elle a fait *le* manger à l'enfant.
b. ?*Elle a fait *en* boire au chat.
c. ?*Il a fait *le* signer à la patronne.

Goodall (1984) argues that the peculiar behavior of this construction with respect to cliticization and other phenomena is not the result of the interaction of a number of specific rules. The peculiarities follow, rather, from a specific phrase structure representation. Goodall's analysis is based on the existence of "parallel simultaneous structures" in the grammar. In the case of coordinated structures, a given phrase marker can contain two distinct terminal strings. The causative construction would constitute an example of a terminal string with more that one structural analysis.

According to Goodall (1984, 144), the sentence in (90a) has the two simultaneous structures given in (98).[33]

(98)a.

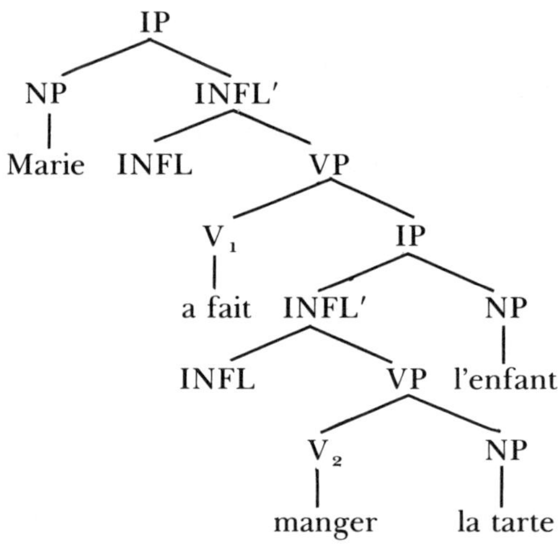

b.

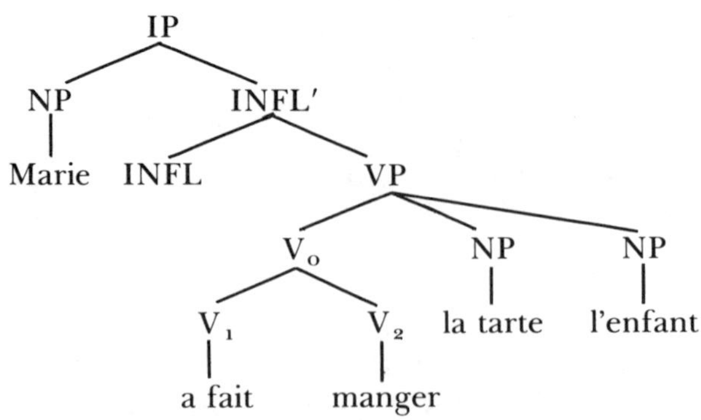

Zubizarreta (1982b) argues that *faire* must satisfy two subcategorization frames: ____ S(or IP) and ____ V. This particular frame is satisfied in the parallel structures in (98). Notice that the NP subject in (98a) is in a postverbal position, contrary to the position it normally occupies in tensed sentences as in (99).

(99)

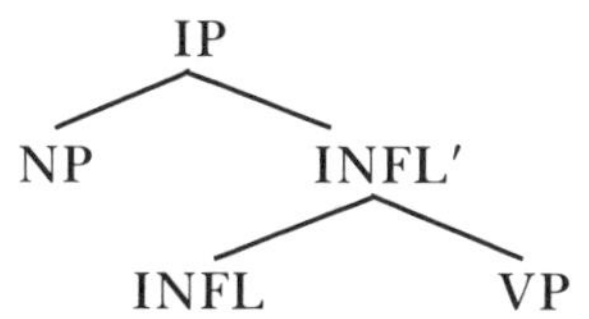

Goodall adopts Pesetsky's (1982) proposal that the order in (99) is required by adjacency requirements on (nominative) Case-assignment. That is, even though INFL governs the external argument *l'enfant* in (98a), there is no need for the two to be adjacent since INFL here is [−TNS] and does not assign nominative Case to the subject. In (99), on the other hand, INFL does assign Case to the subject since it is [+TNS]. Consequently, the NP subject must be adjacent to INFL.

I agree with Goodall that the structures in (98), especially (98b), can provide a plausible account for the peculiar behavior of clitics in causatives without further stipulation. Goodall (1984, 150) makes two assumptions with respect to Case assignment: Case is assigned optionally with the undesirable results of this being ruled out by the Case filter and an NP may not be Case-marked more than once.

Thus, V_2 in (98a) and V_0 in (98b) can assign Case to the object, but if both do so then the NP *la tarte* ends up Case-marked with more than one Case. This is impossible, according to the second assumption. V_0 assigns accusative Case to the NP object and dative Case to the postverbal subject since neither V_1 nor V_2 govern this NP subject in (98). V_0 assigns the Cases belonging to *faire* (____ ACC (DAT)), and not the one belonging to *manger* (____ ACC). This must be true since *manger* has only one Case to assign but there are two NPs in (98b); *l'enfant* would not receive Case. The postverbal subject then surfaces with the dative *à*.[34]

(100) Marie a fait manger la tarte *à* l'enfant.

We can now concentrate our attention on clitics. Consider first the contrasts in (101).

(101)a. *Marie a fait la manger à l'enfant.
b. Marie l'a fait manger à l'enfant.

The structure corresponding to (98b) for (101b) is the one in (102).

(102)

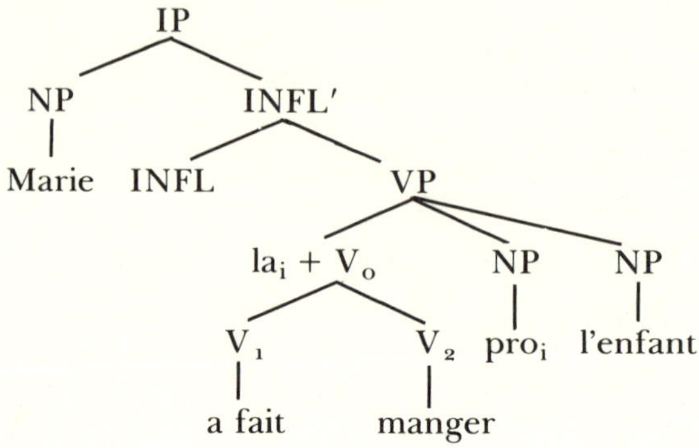

According to what we have said concerning object clitics up to this point, the clitic *la* cannot appear anywhere else in (101b). More precisely, affixation to V_0 is the only possible position for this object clitic for two reasons. First, since the clitic is a Case-absorber it is only natural that *la* be affixed to V_0. Secondly, a clitic has to govern a coindexed pro in argument position in order to license it. If *la* were affixed to V_1 or V_2, this condition would not be met.

The result of this appears on the surface to be an affixation to the higher verb but it is, in fact, an affixation to the verb complex $[_V V_1 V_2]$.

Consider now the same contrast with cliticization of the dative object in (103) and the corresponding structure in (104).

(103)a. *Marie a fait lui manger la tarte.
b. Marie lui a fait manger la tarte.

(104)

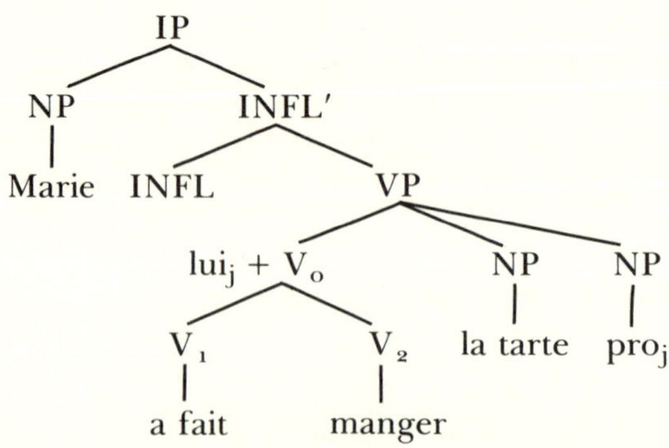

The analysis above holds of the dative clitic appearing on V_0. When both objects are cliticized, the predicted string is the one given in (105) which is the right result.[35]

(105) Marie la_i lui_j a fait manger pro_i pro_j.

This is essentially Goodall's approach adapted to our analysis of object clitics. It can be extended to clitics in Spanish causatives; cf. Goodall (1984, 184-6) and Aissen and Perlmutter (1976) for a similar analysis of Spanish causatives in a Relational Grammar framework and without "parallel structures".

In conclusion, the fact that a clitic corresponding to an object of the lower verb surfaces as an affix on the higher verb in causative constructions is not a counterexample to the claims in 2.2.3 if we adopt Goodall's (1984) views on this construction.[36]

2.6. SUMMARY

In this chapter, I have been principally concerned with the analysis of subject clitics in syntactic structures where they are not doubled by an overt NP in the argument position to which they are related. The hypothesis that subject clitics represent a grammatically real notion and should consequently be analyzed in the same way in all languages exhibiting them was defended on the basis of the following facts.

First, it was noted that in languages with subject clitics simple tensed sentences always have a phonologically realized subject. There is always either a subject clitic or an overt NP or both. This was argued to follow without stipulation from the hypothesis that subject clitics are generated in INFL and serve as licensing elements for pro in external argument position. If the [NP,IP] position is empty then it can only be pro and a clitic must appear in INFL in order to license this occurrence of pro.

Second, such an analysis leads to a principled characterization of the symmetry holding between subject clitics and object clitics with respect to their capacity to license pro in argument position. The two types of clitic can therefore be conceived of as two distinct realizations of the same syntactic element "clitic".

I then suggested an approach to clitics in which they are analyzed as having no status with respect to A-binding or $\bar{A}$-binding. As affixes, they are neither in an A-position nor an $\bar{A}$-position. Furthermore, it was tentatively assumed, following others, that clitics absorb the Case normally assigned by the head they are affixed to.

This section was followed by a more general discussion of the null subject parameter which was argued to be non-existent because pro, like any other empty category, is initially available in UG due to Binding theory. In conformity with the Recoverability Principle, the task of the language-learner consists of determining the conditions under which pro is licensed in the language.

I then turned to related issues, namely, free inversion and causative constructions and argued that the analysis of clitics presented here can handle the apparently problematic facts found in these constructions.

In chapter 3, I discuss how the general approach outlined in this chapter can account for the doubling phenomena for both external and internal arguments.

CHAPTER THREE

On Clitic Doubling

3.0. INTRODUCTION

In this chapter I will be concerned with representation 2(1). I wish to suggest a unified account of the subject doubling construction and point out its similarities and differences with respect to object doubling (3.1 and 3.2).[1] The discussion on subject doubling is based mostly on the Northern Italian dialects and some French dialects. I want to argue that it is not desirable to establish the connection implicitly assumed in Safir (1985) between subject doubling and free inversion. Rather it seems preferable to assume the theory developed in the previous chapter, in which subject clitics are generated in the AGR element of INFL and are associated with pro in external argument position. The possibility of subject doubling is then related to a parameter making reference to the optionality of Case-absorption by the clitic.

I then turn in 3.3 to a major problem related to doubling constructions, namely extractions. An examination of the facts concerning extractions both at LF as well as in syntax will lead to an explicit formulation of the relationship between clitics and arguments. I propose that these elements form what I will call c-chains, which are different from A- or $\bar{A}$-chains. This proposal is consistent with the theory developed so far.

I then examine two related issues in 3.4 and 3.5. First, I wish to argue that the Complex Inversion found in St.Fr. and some N.It. dialects does not involve subject doubling even though the two constructions are very similar on the surface. This chapter ends in 3.5 with a note on the reflexive clitic *se* and, more precisely, how it is, not a clitic and thus does not form a c-chain.

3.1. SUBJECT DOUBLING

3.1.1. Northern Italian Dialects

The facts about the N.It. dialects have been presented in Brandi and Cordin (1981) and Renzi and Vanelli (1982). Different analyses have since been proposed; cf. especially Bouchard (1982), Jaeggli (1984), Rizzi (1986a), and Safir (1985). Except where otherwise mentioned, the data I will use is from Trentino and is taken from the references above.

The question of determining what kind of lexical NP can be doubled by a subject clitic is deferred to 3.3. Subject clitics are obligatory in Trentino whether a lexical preverbal NP is present or not, as in (1), but are excluded in the cases of free inversion as in (2).[2]

(1)a. *(El) magna.
CL eat-3S
'He eats.'
b. El Mario *(el) magna.
the
'Mario eats.'
c. *(Le) ven.
CL come-3P
'They come.'
d. Le putele *(le) ven.
the girls
'The girls come.'

(2)a. (*El) magna el Mario.
b. (*Le) ven le putele.
c. E' vegnu 'na putela.
has come a girl
'A girl came.'
d. *L' ei vegnuda una putela.
CL

As in St.Fr. the subject clitic can be inverted in questions, as shown in (3); cf. 3.4 for the details.

(3) (El Mario) magne*lo*?

This inversion is possible in matrix sentences but excluded from embedded contexts, as shown in (4).

(4)a. Ho domanda se (el Mario) el magna a casa.
'I wondered if (Mario) he eats at home.'
b. *Ho domanda se (el Mario) magnelo a casa.

A sentence like the one in (5), where the preverbal clitic is interpreted as a 3s subject, is ungrammatical. This is not the case if *el* is the homophonous object clitic.

(5) *El magnelo?

These facts suggest the following interpretations. First, Trentino and the other N.It. dialects are null subject languages in the sense defined in chapter 2, because the positions occupied by the clitic and the one occupied by the lexical NP subject cannot generally both be empty in a tensed clause at D-structure.

(6)a. *Ven.
come-3s
'He/she comes.'
b. El/la ven.
'He/she come.'
c. *Magna.
eat-3s
'He/she eats.'
d. El/la magna.
'He/she eats.'

Second, the free inversion property characteristic of Italian and Spanish is found and can be analyzed in the same way as argued in 2.4. Finally, subject doubling is possible, and even obligatory, in some constructions such as in (1) above.

3.1.2. Previous Analyses

The first analysis to be reviewed here is that of Safir (1985). The basic assumptions adopted by Safir in order to account for the doubling phenomenon in (1b,d) and (4a) are that nominative Case must be assigned in VP and that subject clitics must be assigned Case directly. Subject clitics are base generated as affixes on the verb. Therefore, in the structure in (7) corresponding to (1b,d), Case is assigned by INFL to the subject clitic generated inside VP and the lexical NP subject can inherit Case from this coindexed clitic[3] (→ Case assignment; ⇒ Case inheritance).

(7)

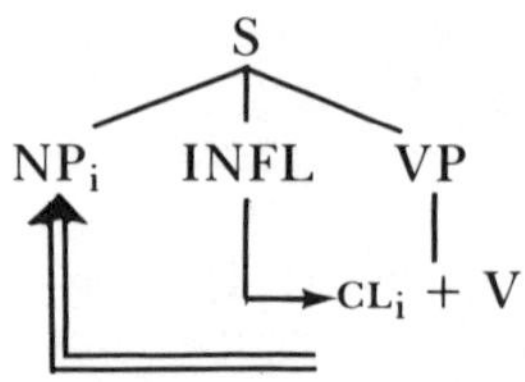

These assumptions seem to account for why the subject clitic is obligatory in non-inverted clauses and how the lexical NP receives nominative Case in subject doubling constructions. Furthermore, the absence of an overt subject clitic in free inversion constructions, as in (2) above (cf. example (83) in chapter 2), derives from Case-assignment through government by the impersonal clitic to the inverted NP as in (8). That is because Case assignment is the transmission of a feature which renders the subject clitic "incomplete" and thus phonologically null (→ Case-assignment).

(8)

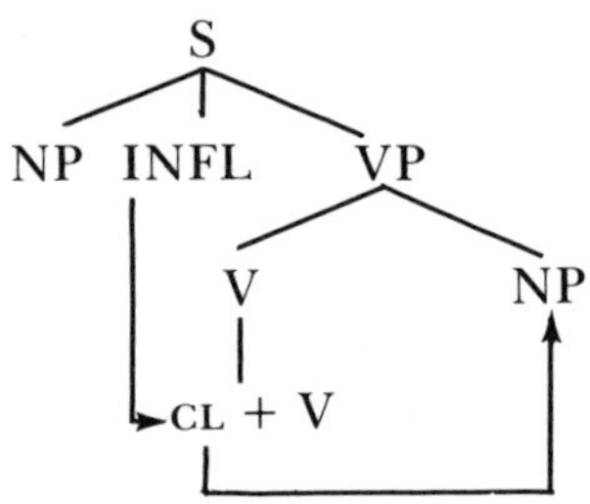

Case inheritance in free inversion constructions is impossible since the clitic and the NP are not coindexed. Thus, the constructions in (7) and (8) differ in that in the former the clitic and the NP share a Case whereas in the latter the NP alone is Case-marked by INFL through the clitic. From this is derived the fact that there is no definiteness effect, i.e. there is no Θ-chain formed; for a detailed account of the definiteness effect see Safir (1985).

This analysis of free inversion resembles the one adopted in Chomsky (1981) in that nominative Case being in VP now can be assigned under government to the inverted NP; cf. also 2.4.

Note finally that subject doubling and free inversion are connected in this approach in the sense that, for an inverted NP to be realized phonetically, the VP must contain a clitic which has been assigned nominative Case directly by INFL. Two things may then happen: the clitic retains its Case, resulting in the sentence in (6b,d) or the clitic assigns Case to the inverted NP, resulting in free inversion.

There is also the possibility that an NP in subject position may inherit Case from the clitic. The result in this case is a subject doubling construction. Thus, as long as there is a subject clitic assigned Case in VP, all the conditions on subject doubling and free inversion are met. Free inversion is made possible by the parameter in (9).

(9) Free Inversion Parameter
If the (impersonal) subject clitic is assigned
Case by INFL, then it can, in turn, assign Case.
(Safir 1985, 234)

For languages in which there is no subject doubling, for instance Standard Italian and Standard French, one must stipulate that there is no Case-inheritance. The NP subject then does not get Case and does not pass the Case Filter. It follows that only languages with subject clitics can have free inversion since free inversion is made possible only by the assignment of nominative Case to the inverted subject by a subject clitic affixed to V; Safir (1985, 266). It is a well known fact that St.It. has no (overt) subject clitics but has free inversion. Thus Safir (1985, 259) makes the following claim:

(10) If a language L has any argument clitics (subject, object, indirect object) associated with clitic slots on a verb, then, in the unmarked case, L has the full clitic paradigm.

This claim allows him to assume that St.It. has silent subject clitics. Note that this assumption is incompatible with my speculations concerning verbal agreement in inversion constructions; cf. example (89) in chapter 2. Moreover, the analysis of free inversion suggested in 2.4 does not require one to claim something like (10), which I take to be an unfalsifiable claim. I will therefore not assume it in the rest of this book.

The second analysis is found in Rizzi (1986a). Recall that Rizzi adopts the structure in (11) for sentences like the ones in (1b,d).

(11)

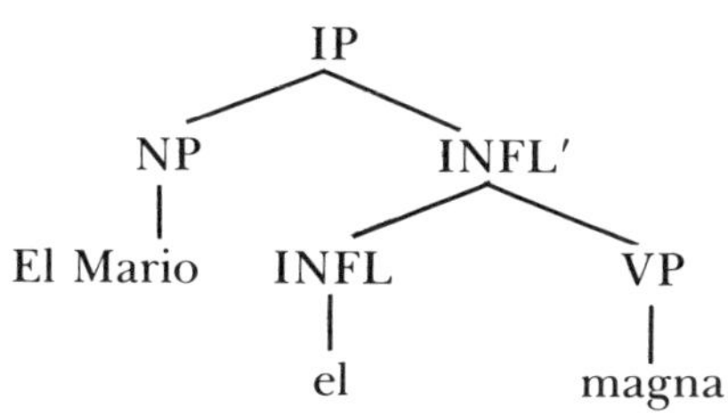

Of course the representation in (12) is also available in the N.It. dialects.

(12)

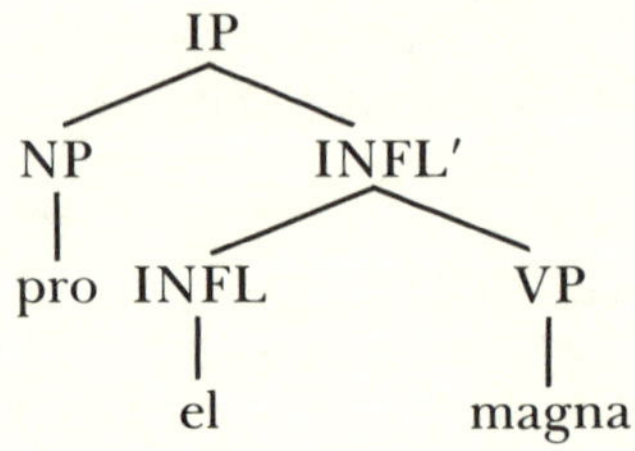

Rizzi (1986a, 393) adopts an analysis in terms of "reduplication of agreement" and claims that the N.It. dialects only differ from St.It. in that "the strong AGR is realized in Phonetic Form not only in the concrete verbal morphology, but also in its abstract syntactic position [INFL]." There is no commitment made as far as Case and Θ-roles are concerned although Rizzi offers two alternatives. The first alternative is that the clitic is a true nominal and is thus not an argument (given the presence of a lexical or empty argument in [NP,IP]); the clitic and the NP subject form a syntactic chain and share the same Case and Θ-role. The second alternative is that the clitic is not a true nominal; its presence is therefore irrelevant for Case assignment and Θ-role assignment.

This analysis leaves a certain number of important questions unanswered. How are the apparent intermediate cases represented by Colloquial French and Pied Noir French analyzed? Subject doubling is more restricted in these French dialects than in the N.It. dialects. If subject clitics are truly just another morphological realization of AGR, how can the striking similarities between subject clitics and object clitics pointed out in 2.2.3 be accounted for? Why are subject doubling and object doubling analyzed as two completely distinct phenomena given that they are so similar in many respects (cf. 3.2)? Why would a simple morphological realization of AGR have the influence on extraction possibilities that it apparently has (cf. 3.3)?

I will argue in the following sections that answers to these questions can be provided by a theory that makes specific proposals concerning the properties of subject clitics with respect to Case theory.

3.1.3. French Dialects

3.1.3.1. Colloquial French. As mentioned in 2.1, we will be concerned with two French dialects in addition to the N.It. dialects. The first dialect is Colloquial French, which exhibits a variation very similar to the property of the N.It. dialects with respect to the distribution of lexical NPs and subject clitics.

In this dialect a lexical NP can optionally be doubled by a referential subject clitic as in (13).[4]

(13)a. Marie (elle) vient.
'Marie (she) comes.'
b. Pierre (il) mange.
'Pierre (he) eats.'
c. Cet homme (il) a dit qu'il allait embrasser M. Caouette.
'This man (he) said that he would kiss Mr. Caouette.'
d. Les loyers (ils) ont augmenté.
'The rents (they) have increased.'

Now, it is not obvious that the above examples are instances of doubling and not left-dislocations of the subject, for which the appropriate structure would be the one in (14) argued for in St.Fr. by Rizzi (1986a) with cliticization to the right in PF; cf. Kayne (1984) and 2.2.4.

(14)

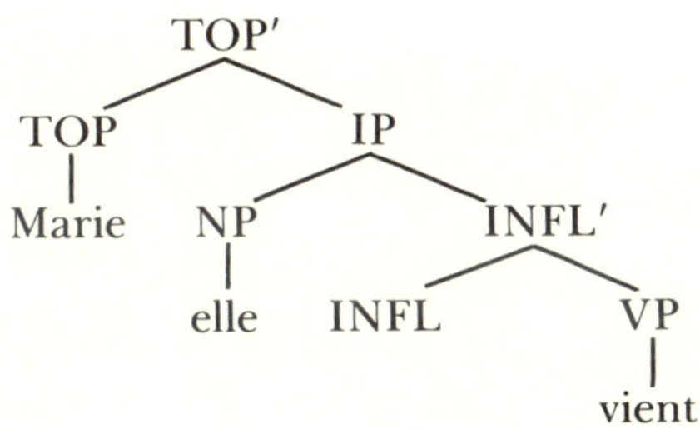

Carroll (1982) provides some tests to distinguish dislocations from doubling constructions. Interestingly, of all the examples she examines, only those similar to the ones in (13) are not decisively cases of (left-)dislocations according to her tests. Some examples of dislocations are provided in (15).

(15)a. Il est parti, Jean.
'He is gone, Jean.'
b. Pierre est fin, lui.
'Pierre is nice, him.'
c. Dodo est malade, la pauvre.
'Dodo is sick, the poor (girl).'

That the sentences above are dislocations is determined on the basis of various tests. For example, there is a strong pause between the dislocated constituent and the rest of the sentence.[5] A dislocated constituent does not occupy a position that receives a grammatical function. The ungrammaticality of the sentences in (16) would thus be explained indirectly by the lack of an element to fill this function.

(16)a. *Est parti, Jean.
b. *Est malade, la pauvre.

On the other hand, in instances of doubling there is no obligatory pause between the two elements involved. This is the case in the examples in (13). Furthermore, the elements involved in doubling constructions are in "regular" syntactic positions. It should therefore be the case that there is less freedom in the choice of these constituents. A doubled NP subject, for instance, could be doubled only by a subject clitic which is in INFL, a syntactic position available in D-structure, as argued in chapter 2. This is in fact what happens. It is impossible to find *doubling* constructions like the ones in (17), compare to (15).

(17)a. *Pierre lui est fin.
b. *Dodo la pauvre est malade.

Only subject clitics can intervene between a lexical NP subject and a verb as in (13). If the structure of the sentences in (13) were that in (14) there would be no reason to exclude any type of NP from appearing in [NP,IP]. On the other hand, with a structure like (11), this fact receives a natural explanation: the leftmost NP occupies the external argument position, making it impossible for other NPs like *lui* and *la pauvre* in (17) to intervene. The only syntactic position available is INFL where subject clitics are generated. The contrast between (13) and (17) is thus accounted for.

Note that the sentences in (17) have grammatical equivalents in Standard French and in Colloquial French. In the former, an obligatory pause is made between the first NP and the second NP as in (18a,b) whereas, in the latter, a subject clitic appears together with the pause as in (18c,d). These sentences are used in order to emphasize the subject.

(18)a. Pierre, lui, est fin.
b. Dodo, la pauvre, est malade.

c. Pierre, lui, il est fin.
d. Dodo, la pauvre, elle est malade.

Studies have shown that left-dislocations in English always introduce or reintroduce a referent in the discourse. In both cases, the referent can be considered as 'discourse-new'; see in particular Keenan and Schieffelin (1976). It is not the case in C.Fr. that sentences like the ones in (13) with subject clitics involve a discourse-new referent.

Moreover, an object can be left-dislocated in C.Fr. as in (19).

(19)a. *A Whistler*, on *y* va souvent.
'To Whistler, we go there often.'
b. *Jean*, je *le* rencontre de temps en temps.
'Jean, I meet him occasionally.'

When a left-dislocated object appears along with subject doubling, the object is found to the left of the subject.

(20)a. A Whistler, mes amis ils y vont souvent.
'To Whistler, my friends (they) go there often.'
b. Jean, Marie elle le rencontre souvent.
'Jean, Marie (she) meets him often.'

An interpretation in which both the object and the subject are left-dislocated is possible, either by making a strong pause (##) between the subject *Marie* and the clitic *elle* or by moving the subject further to the left, in front of the lexical object as in (21b,c).

(21)a. Pierre$_i$, Marie$_j$ elle$_j$ l$_i$'a vu.
'Pierre, Marie (she) has seen him.'
b. Pierre, Marie ## elle l'a vu.
c. Marie, Pierre, elle l'a vu.

A sentence like (21a) with regular phonological phrasing, i.e. without a strong pause, cannot be interpreted as involving a left-dislocation of the subject in C.Fr.

If one accepts the structure in (14) for the sentences in (13), then one would expect the leftmost NP to appear to the left of the material found in COMP or the Specifier of COMP, namely complementizers and Wh-phrases. This is so since the more explicit version of the structure in (14) is that in (22), following Chomsky (1977).

(22)

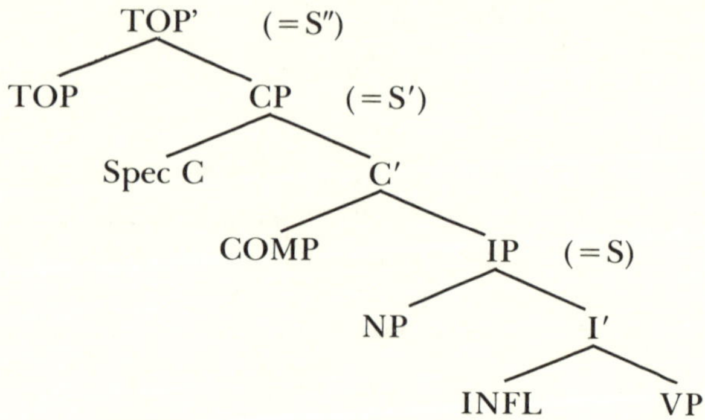

The facts do not accord with this prediction. In C.Fr., when the construction is used in embedded contexts, the lexical NP is found to the right of the complementizer or *wh*-phrase, not to the left as (22) would predict.

(23)a. Il a fallu que maman elle aille travailler sur
it was needed that mom she go to + work on
la ferme.
the farm
'It was necessary for mom to go work at the farm.'

b. Je me souviens quand la petite fille du
I REFL recall when the little girl of + the
voisin elle s' est noyée.
neighbour she REFL is drowned
'I remember when the neighbour's daughter drowned.'

The same is true when the construction is used in matrix sentences.

(24)a. Qui que Marie elle veut voir?
'Who does Marie want to see?'

b. Quand que Pierre il a été malade?
'When was Pierre sick?'

If the lexical NP is to the left of the material in COMP or the Spec of Comp, as in (25) corresponding to (23) and (24), then, and only then, can it be interpreted as left-dislocated.

(25)a. Il a fallu, maman, qu'elle aille travailler sur la ferme

b. Je me souviens, la petite fille du voisin, quand elle s'est noyée.

c. Marie, qui qu'elle veut voir?

d. Pierre, quand qu'il a été malade?

Or alternatively:

(26)a. Maman, il a fallu qu'elle aille travailler sur la ferme
b. Il a fallu qu'elle aille travailler sur la ferme, maman

(27)a. La petite fille du voisin, je me souviens quand elle s'est noyée.
b. Je me souviens quand elle s'est noyée, la petite fille du voisin.

The NPs *maman* and *la petite fille du voisin* are in dislocated positions in (25), (26) and (27) but not in (23).

Finally, there is a constraint on the nature of the lexical NP used in sentences like (13): the NP cannot be indefinite (or non-specific), quantified or [+wh] as shown in (28)-(30).

(28)a. Son ami il est toujours là.
'His friend (he) is always there.'
b. *Un ami il est toujours là.
'A friend (he) is always there.'
c. Un ami est toujours là.

(29)a. Mes amis ils viennent à soir.
'My friends (they) come tonight.'
b. *Beaucoup d'amis ils viennent à soir.
'A lot of friends (they) come tonight.'
c. Beaucoup d'amis viennent à soir.

(30)a. Pierre il aime la tarte.
'Pierre (he) likes pies.'
b. *Qui il aime la tarte?
who he likes the pie
'Who likes pie?'
c. Qui aime la tarte?

This particular constraint is often used to argue against an analysis of this construction in terms of subject doubling and for one in terms of left-dislocation; cf. Galambos (1980) and Rizzi (1986a). This conclusion is problematic, however, and these facts can be interpreted very differently since this constraint is often associated with object doubling constructions; cf. Borer (1984), Jaeggli (1982), and Steriade (1980) among others. The River Plate (henceforth, RP) Spanish examples in (31)-(33) illustrate this.

(31) *Lo_i ví a un $chico_i$.
him saw to a boy
'I saw a boy.'

(32) *Las$_i$ ví a todas las chicas$_i$.
them saw to all the girls
'I saw all the girls.'

(33) *A quién$_i$ lo$_i$ vimos t_i?
to whom him saw
'Who did we see?'

However, object doubling is grammatical with a definite object as in (34) and extraction of the object can take place only when it is not doubled by a clitic as in (35).

(34) Lo$_i$ vimos a Guille$_i$.
him saw to
'We saw Guille.'

(35) A quien$_i$ vimos t_i?

A detailed discussion of this particular constraint on the type of NP that can be clitic doubled is presented in 3.3. What is important to note for the moment is that such a constraint cannot be used to argue definitively against a doubling analysis of the sentences in (13) and that it can, in fact, support such an analysis in the light of the similar property found in "attested" object-doubling constructions like (34).

Thus, I conclude that the C.Fr. sentences in (13), (18), (20), (21a), (23), (24), (28a), (29a), and (30a) are all instances of subject doubling and that an obligatory left-dislocation interpretation necessitates appeal to further modifications of the structure as in (15), (21b,c), (25), (26), or (27).[6]

The structure associated with subject doubling in C.Fr. is the one in (11), repeated in (36). This structure is consistent with the analysis of subject clitics introduced in chapter 2.

(36)

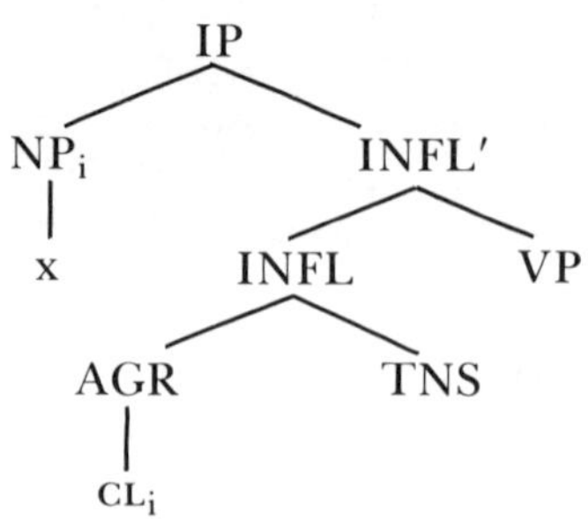

With respect to this conclusion, it is interesting to note that Vanelli and Renzi (1982, 134) acknowledge the existence of subject doubling in C.Fr.; they write: "Anche il francese presenta simili varieta [NP-pronoun-verb] (...); e cosi si comporterebbe anche il français populaire" for which they refer to Guiraud (1973) and Meyer-Lübke (1889).

3.1.3.2. Pied Noir French. The second French dialect to be considered here is the one often referred to as Pied Noir French; cf. footnote 3 in chapter 2.[7]

The situation in this dialect is rather simple in light of the facts already introduced for C. French. PN French has subject doubling. This is illustrated in (37).

(37)a. L'homme (il) vient.
'The man (he) comes.'
b. Le soleil (il) brille pour tout le monde.
'The sun (it) shines for everybody.'

This doubling is optional as it is in C.Fr. but it is close to the N.It. dialects in that any type of NP can be doubled. Example (38c) is from Brua (1938, 11).

(38)a. Un homme il vient.
'A man (he) comes.'
b. Chaque femme elle parle.
'Each woman (she) speaks.'
c. Personne il sait qui c'est leur mère.
'Nobody (he) knows who their mother is.'

In PN French, there is also object doubling as in (39).[8]

(39)a. Marie l_i' aime à $Jean_i$.
him likes to
'Marie likes Jean.'
b. La fille l_i' a vu à $Jean_i$.
the girl him has seen to
'The girl has seen Jean.'

There is also clitic doubling with indirect objects as in (40).

(40)a. Je lui_i ai donné la pomme au $professeur_i$.
'I gave the apple to the teacher.'

b. Tu lui_i as donné des fleurs à $Marie_i$.
'You gave flowers to Marie.'

The object-doubling property of PN Fr. will be discussed in more detail in 3.2.

3.1.4. Analysis

I now turn to the specific analysis of subject doubling. The crucial fact that will guide my analysis is the impossibility of free inversion in C.Fr. as shown in (41).

(41)a. *Est parti Jean.
is gone
'Jean is gone.'
b. *Arrive Marie.
'Marie arrives.'

Recall that Safir's (1985) account of the subject-doubling property in Trentino and the other N.It. dialects is influenced by the free inversion construction these languages exhibit. When generating the subject clitic inside VP and assigning Case to it directly, it seems natural to Safir to assume that this subject clitic can, in turn, assign Case to an inverted NP adjoined to and directly dominated by VP since, under a particular definition of c-command, the subject clitic governs the inverted NP.[9]

On the other hand, this account of subject doubling requires formation of a chain (coindexation) between the clitic and the preverbal NP and inheritance of Case; cf. 3.1.2.

There is no free inversion in C.Fr.. It should thus be possible to suggest an analysis that does not require an empty subject clitic in VP in order for the preverbal NP to inherit Case and to assign Case to a postverbal NP. Recall also that this type of analysis leads to the postulation of (silent) subject clitics in languages with free inversion and no overt subject clitics such as St.It. and Spanish.

In 2.4, I have proposed an analysis of free inversion that does not require empty preverbal subject clitics; the postverbal subject can get nominative Case directly within VP. In 2.2.5, it was suggested that clitics in general absorb Case. The following account is based on these premises.

Let us assume, as we have done so far, that in C.Fr. the subject clitic is base generated under AGR inside INFL as in (42).

(42)

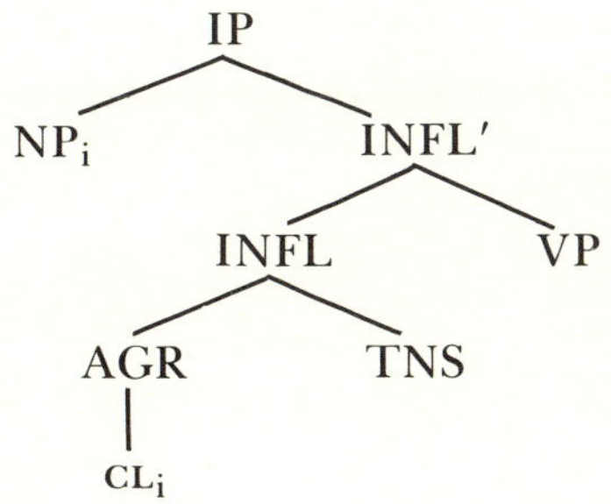

Given the Case Filter, the question that remains concerns Case assignment to the NP when it is lexically realized. Given that nominative Case is assigned by the (+)TNS element, I assume the following parameter, to be modified later on.

(43) A subject clitic need not absorb Case.

If the clitic in (42) absorbs Case, then the NP cannot have lexical content; such an NP would be excluded by the Case Filter. In order for a lexical NP to be realized phonetically there must be no subject clitic in AGR. This is the situation in Standard French, as shown in (44).

(44)a. Jean vient.
'Jean comes.'
b. Il vient.
'He comes.'
c. *Jean il vient.
d. *Vient.

If the clitic does not absorb Case, then this Case can and must be assigned to the NP, which would therefore be lexical. This is so even under an approach to Case assignment in which this process is optional. Suppose that Case is not assigned at all, then no lexical element can appear in preverbal position, be it a full NP or a subject clitic. What, then, would be the empty category appearing in [NP,IP]? It cannot be a trace since the position is not properly governed and it cannot be PRO since it is governed. It can only be pro, but then this occurrence of pro is not licensed because this is done only by a clitic in French. Therefore, when a clitic is generated, either pro or a lexical NP can appear in [NP,IP]. Of course, in the case where no subject clitic is generated in AGR, only the NP is realized.

We can account in this manner for the optional status of subject doubling in Colloquial French shown in (45).

(45)a. Jean vient.
b. Il vient.
c. Jean il vient.
d. *Vient.

In summary, the difference between C.Fr. and St.Fr. is parametrized as in (43). In St.Fr., (43) is fixed negatively (subject clitics always absorb Case) on the basis of the fact that a subject clitic always appears without an associated lexical NP in preverbal position. In C.Fr. on the other hand, since the co-occurrence of a subject clitic and a preverbal lexical NP subject is attested, (43) is fixed positively.

PN French patterns with C.Fr. in the setting of the parameter in (43). The differences between them with respect to the type of NP that can be doubled will be explored in 3.3.

With respect to learnability, it must be assumed that the unmarked situation is the one in which the clitic absorbs Case. Contact with subject-doubling constructions is sufficient for the positive setting of the parameter in (43).

The analysis presented above, which readily accounts for the optionality of subject doubling in C. French and PN French and its impossibility in St. French, can be extended to the N.It. dialects that have subject doubling such as Trentino and Fiorentino.

The differences between the N.It. dialects and the French dialects are clear. In Trentino, subject doubling is obligatory and there is free inversion. The situation is the following. In Trentino when nominative Case is assigned preverbally, the subject clitics must be realized phonetically, as shown in (1), whereas when it is assigned postverbally; the subject clitics do not surface, as shown in (2). It is therefore possible to assume that in Trentino a subject clitic is always generated at D-structure in AGR. Accordingly, the parameter in (43) is set positively so that the preverbal lexical NP, when present, can be assigned Case. This also holds in Fiorentino.

Now, subject clitics in Trentino are excluded in the free inversion construction but in Fiorentino, when free inversion occurs, the neuter clitic *e* appears in preverbal position as in (46); cf. also 2.4.

(46) E' viene le ragazze.
it come-3s the girls
'The girls come.'

Crucially, the verb agrees with the pro in [NP,IP] which is licensed by the neuter clitic, and not with the inverted NP. This is consistent with the approach developed here. Consider the S-structure in (47) corresponding to (46).

(47)

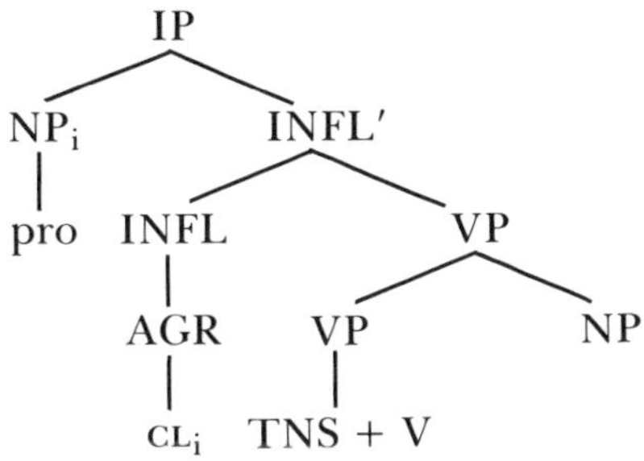

In (47), TNS assigns nominative Case to the postverbal NP, the clitic does not absorb Case in compliance with (43) and it licenses pro in [NP,IP]. The preverbal NP subject position is not Θ-marked, so pro is expletive and an expletive (or neuter, or impersonal) clitic appears.

In Trentino, we can simply assume, following Brandi and Cordin (1981) and Rizzi (1986a), that there is a gap in the clitic paradigm corresponding to the neuter clitic (among others). This assumption is supported by the fact that the verb does not agree with the inverted NP as it does in Fiorentino. This is illustrated by the contrast in (48).

(48)a. E' vegnu 'na putela.
has come a girl
'A girl came.'
b. *L'ei vegnuda 'na putela.

In (48a) then, as in all other instances of free inversion in Trentino, no preverbal subject clitic can be generated since none is available in the grammar of this dialect. The verb thus agrees with the set of features in the preverbal position that would be realized as a clitic (as in Fiorentino) if there were a clitic corresponding to those features in Trentino. Pro in this case is licensed by this set of features as it always is in Standard Italian. This set of features can be considered to be an empty clitic.

This last point brings up the delicate issue raised by the postulation of empty (or silent) clitics. Recall that in 3.1.2 above I observed that Safir's (1985) assumption concerning the presence in Standard Italian of silent clitics was incompatible with the approach developed here. However, I believe that there is a significant difference between Safir's assumption and the one presented here. This difference has to do with the concept of learnability. Suppose that a language learner is exposed to some St.It. sentences. Of course, none of these sentences contains subject clitics. Therefore, in order for the learner to "learn" the existence of the silent subject clitic, it must resort to UG which, according to Safir, includes a statement such as the one in (10) above. The trigger experience in this case is the presence of object clitics in the available data.

The situation is quite different in the instances of gaps in the paradigm of clitics as in the case of the subject clitic paradigm in Trentino. In this particular instance, the language learner is exposed to concrete evidence that there are subject clitics in the grammar and, also, that, for specific sets of features in INFL, there are no phonetic realizations corresponding to a clitic. There is thus no need here for a statement like the the one in (10).

In this respect my position is closer to the one proposed in Rizzi (1986a, 402):

> In the [N.It.] dialects the subject clitic is the strong AGR licensing pro in a N[ull] S[ubject] L[anguage], and these grammatical systems differ from Standard Italian only in that the AGR specification receives phonetic content also in its abstract syntactic position. It is natural to think of this trivial parameter as possibly assuming different values for different grammatical specifications of AGR; i.e., the phonetic spell-out of AGR under INFL is obligatory or optional, or impossible for different specifications of person and number in the [N.It.] dialects, and always impossible in Standard Italian.

I disagree with Rizzi in his interpretation of where French fits within the above description of the facts; cf. 3.1.2.

To summarize, in my account of Trentino, a subject clitic is always generated in D-structure in AGR and (43) is set positively – as it is in C.Fr. – but a major difference between C.Fr. and Trentino follows from the possibility of free inversion in the latter. Base generation of the NP subject in postverbal position accounts for the fact that there is no agreement between the verb and this postverbal NP in free inversion constructions and this follows from the presence of subject clitics in this dialect. The verb agrees with expletive pro in external argument position which is licensed by a neuter (or impersonal) clitic in Fiorentino or a set of features in Trentino.

The differences between C.Fr. and PN Fr., on one hand, and St.Fr., on the other hand, follow from a difference in the setting of the parameter in (43). C.Fr. and PN Fr. are distinguished from

Trentino and Fiorentino by the possibility of free inversion and obligatory base generation of subject clitics. C.Fr. differs from PN French, Trentino, and Fiorentino with respect to extraction possibilities, a topic to be explored in 3.3.

3.2. OBJECT DOUBLING AND SUBJECT DOUBLING

In 2.2.3, it was argued that the theory of subject clitics developed here also covers the constructions with object clitics in a variety of languages. This is made clear if we adopt the structure in (41) in chapter 2, repeated in (49).

(49)

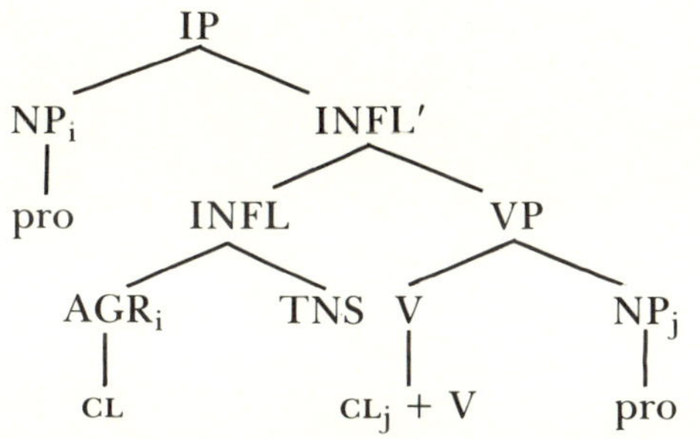

More precisely, it can be derived from the Recoverability Principle that an object clitic must be affixed to the verb when there is a base generated empty category in object position. The object position is coindexed with and governed by the clitic – in fact, the verb – which, in turn, licenses the empty pronominal.

There is thus one major subject/object symmetry as far as the licensing of an empty argument is concerned, namely that this process operates under the same structural condition in both cases.

The goal of this section is to explore subject/object symmetries and asymmetries in the relationship that a clitic has with a lexical argument. In other words, I want to apply the type of discussion that can be found in 2.2.3 to clitic-doubling constructions by using the account of subject doubling introduced in 3.1 and applying it to object-doubling constructions. The rationale behind this should be clear. I have already argued that subject clitics and object clitics are two distinct realizations of the same syntactic element, the function of which is to license pro in argument position. Consequently, the licensing relationship between a clitic and the argument position it is associated with is always the same in the relevant respects whether the argument position is internal or external. This leads to the question explored here: is the relationship between a doubled lexical argument and its associated clitic the same for internal and external clitic-doubled lexical argument?

In the Government and Binding framework, the predicted answer is positive with one qualification. If there are differences, these differences should be explainable in terms of already known subject/object asymmetries, especially with respect to government of the argument positions involved.

3.2.1. *Structures*

I now turn to the structures involved in the two constructions and then turn to the problem of Case-assignment. If the structure in (49) is a plausible one and if the structure for subject doubling is as shown in (50), where *x* is lexical, then the structure for object doubling should be the one in (51).

(50)

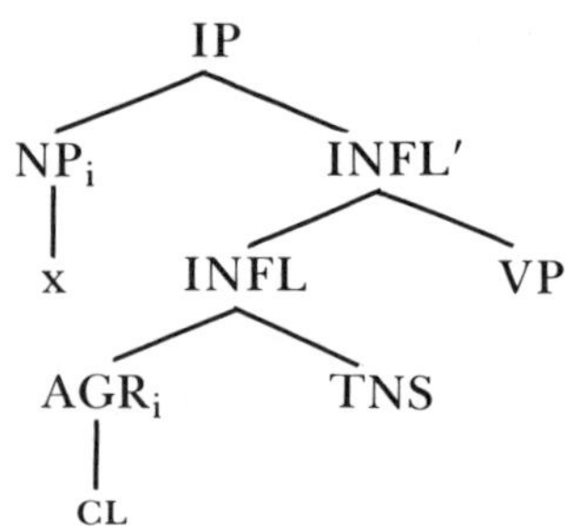

(51)

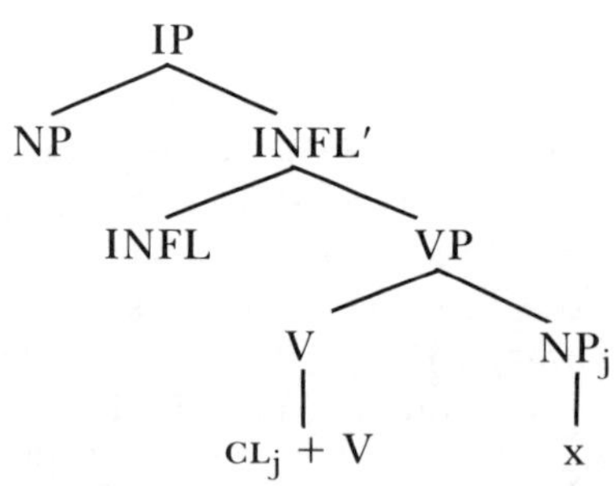

It is an empirical issue to determine whether the structure in (51) is a valid one. The position of the clitic does not raise major problems.[10] In the examples in (52)-(55), the clitics are all affixed to the verb on surface. There is thus no reason to believe that it should not be so at D-structure.

(52) River Plate Spanish
a. Vimos a Juan.
saw-1P
'We saw Juan.'
b. Lo vimos.
CL saw-1P
'We saw him.'
c. Lo vimos a Juan.
CL saw-1P to
'We saw Juan.'

(53) Romanian
a. Am văzut Popescu.
have-1s seen
'I have seen Popescu.'
b. L-am văzut.
CL have-1s seen
'I have seen him.'
c. L-am văzut pe Popescu.
CL have-1s seen OBJ
'I have seen Popescu.'

(54) Pied Noir French
a. Marie aime Jean.
like-3s
'Marie likes Jean.'
b. Marie l' aime.
CL like-3s
'Marie likes him.'
c. Marie l' aime à Jean.
CL like-3s to
'Marie likes Jean.'

(55) Iraqi Arabic
a. Šif-it Xalid.
saw-1s
'I saw Xalid.'
b. Šif-t-a.
saw-1s-CL
'I saw him.'
c. Šif-t-a l -Xalid
saw-1s-CL to
'I saw Xalid.'

More problematic is the position of the lexical NP object in the doubling construction. In the (a) sentences above, the object is, for obvious reasons, in internal argument position [NP,VP] but in the (c) examples this is less evident. Hurtado (1984a;1984b;1985) defends the hypothesis that in the RP Spanish sentence in (52c), the clitic doubled NP *Juan* is in the Ā-position usually occupied by right-dislocated NPs whereas Strozer (1976), Rivas (1977), Jaeggli (1982;1986) and Borer (1984) argue for the hypothesis that this NP is in argument position. I will not review the various arguments here but I accept the second hypothesis: a clitic doubled object NP is in [NP,VP]. The reader is referred to Jaeggli (1986) for some convincing arguments based on intonation, the use of dummy Case-markers (cf. 3.2.2), and bounding effects.

In conclusion, the structure in (51) is a valid one to describe object doubling in the examples (52)-(55) and it is, as predicted, parallel in every respect to the one suggested in 3.1.4 for subject doubling.

3.2.2. Case Assignment

Case-assignment to a clitic doubled NP is at first sight the main source of difference between subject doubling and object doubling.

Following Jaeggli (1982), Chomsky (1982) argues that in a construction like the one in (51) above, the clitic absorbs the Case normally assigned by the verb to its complement. Crucially though, the clitic may or may not manifest this Case and it forms a chain with the NP object. This chain must manifest Case in order to be well-formed according to the Visibility hypothesis. The intuition underlying the Visibility hypothesis is that a "Θ-role is assigned to an element α in a Θ-marked position P and is then inherited [...] by an argument in the chain containing α, but only if the chain has Case" (Chomsky 1981, 334). Thus, the chain (cl,NP) must manifest Case in order to receive a Θ-role. Therefore, if the clitic does not manifest Case, some other member of the chain (cl,NP) must do so. In order for the NP to receive Case, some other means must be used since the verb can no longer accomplish this task because the Case it could assign has already been absorbed by the clitic. This is Chomsky's (1982) way of capturing what is referred to in Jaeggli (1982) as Kayne's gneralization. This generalization states that a clitic-doubled NP always appears with a preposition (a dummy Case-marker).

The particular grammars of River Plate Spanish, Pied Noir French, Romanian, and Iraqi Arabic, for instance, all have recourse to the use of a preposition (the so-called dummy Case-marker) in order to Case-mark the clitic doubled object NPs. This preposition is underlined in (56)-(59).

(56) River Plate Spanish
Lo$_j$ vimos *a* Juan$_j$.
'We saw Juan.'

(57) Pied Noir French
Marie l'$_j$ aime *à* Jean$_j$.
'Marie likes Jean.'

(58) Romanian
L$_j$-am văzut *pe* Popescu$_j$.
'I have seen Popescu.'

(59) Iraqi Arabic
Šif-t-a$_j$ *l*-Xalid$_j$.
'I saw Xalid.'

This process seems to be widespread and is well attested in the literature. Similar processes can be found in other doubling constructions such as the construct state in Hebrew (60) and Iraqi Arabic (61), and object of a preposition in Iraqi Arabic, Tigre and Lebanese Arabic, (62)-(64); cf. also 2.2.1.[11]

(60) Hebrew (Borer 1984)
a. beit -a
house-CL
'her house'
b. beit -a$_j$ *(*šel*) ha -mora$_j$
house-her of the-teacher
'the teacher's house'

(61) Iraqi Arabic
a. ʔumm -a
mother-CL
'his mother'
b. ʔumm -a$_j$ *(*l*)-Xalid$_j$
mother-CL to
'Xalid's mother'

(62) Iraqi Arabic
a. wiya-ha
with-CL
'with her'
b. wiya-ha$_j$ *(*l*)-mara$_j$
with-CL to woman
'with the woman'

(63) Tigre (Jake 1980)
Lilat *(*'ɨgɨl*) ʔali$_j$ warakat nad'at 'ɨt-tu$_j$.
to Ali letter sent to-CL
'Lilat sent a letter to Ali.'

(64) Lebanese Arabic (Aoun 1981)
Ḥkit maʔ -o$_j$ *(*la*)-Karim$_j$.
talked-1S with CL to
'I talked with Karim.'

Now let us consider the following examples of subject doubling in C. French, PN French, Trentino, Fiorentino, and Pirahã.

(65) Colloquial French; Pied Noir French
Jean$_i$ il$_i$ part.
CL leave-3S
'Jean leaves.'

(66) Trentino
El Mario$_i$ el$_i$ magna.
the CL eat-3S
'Mario eats.'

(67) Fiorentino
Le ragazze$_i$ le$_i$ vengano.
the girls CL come-3P
'The girls come.'

(68) Pirahã (Everett 1985)
Kóxoí$_i$ hi$_i$ gi xibáobá.
CL you hit
'Koxoi hit you.'

It is clear from the examples in (65)-(68) that Kayne's generalization does not appear to hold as there are no dummy Case-markers on the doubled NP subjects in these languages.[12] The question we thus want to ask is the following: why is there no dummy Case-marker in subject-doubling constructions?

In order to answer such a question, I will first look at the syntactic structures involved in the four doubling constructions involved here, three of which obey Kayne's generalization. I will then propose a solution based in part on some recent work by Jaeggli (1986) and Levin and Massam (1985).

The structures we are concerned with are: object of a verb, object of a noun, object of a preposition, and subject of a sentence. They are represented in (69) to (72). I adopt Borer's (1984) representations of the first three.[13]

(69)

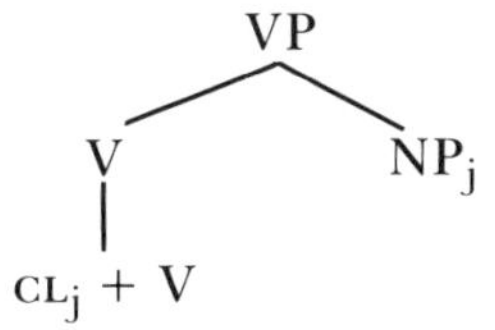

(70)

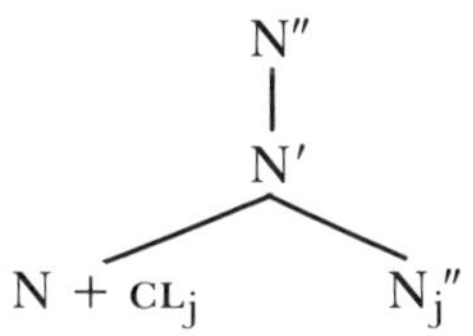

(71)

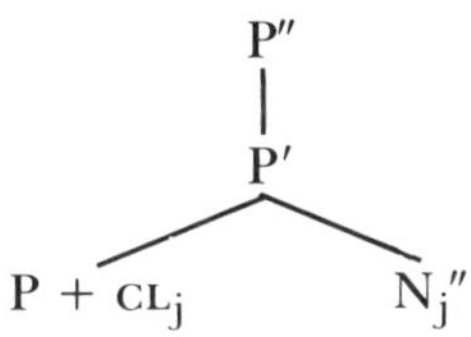

(72)

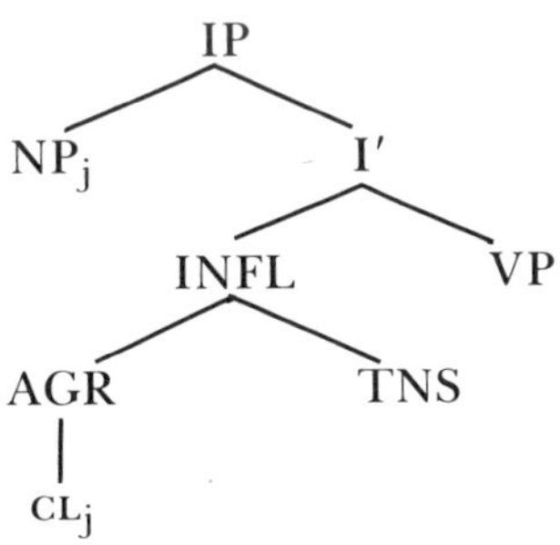

The first major difference – maybe the only relevant one here – is that the three constructions accepting dummy Case-markers in (69), (70) and (71) involve a doubled NP governed by a lexical head. This is not the case in (72).

Before building an account on this observation, it is necessary to examine the exact function and status of dummy Case-markers.

It seems obvious that the dummy Case-Markers are "saving devices," which means that they somehow allow Case to be assigned to a clitic-doubled NP which would otherwise be ruled out by the Case Filter. This is a descriptive statement that provides no account as to the way in which the dummy Case-marker functions. In fact, this element can be either a Case-assigner or a Case-transmitter. The choice of one or the other has important consequences for the status of clitics with respect to Case-absorption. These consequences should, in turn, help in opting for one or the other hypothesis. Let us look at them.

Recall two of my basic assumptions. First, object clitics and subject clitics are just two distinct realizations of a unique syntactic concept. This is based on my observations concerning their function as licensing elements for pro in the argument position they are related to; cf. 2.2 and 2.5. Secondly, given the facts about subject doubling, I have suggested that subject clitics absorb nominative Case (2.2.5) but that this is a parametrized option (3.1.4). In languages allowing subject doubling, subject clitics need not absorb Case and this Case can freely be assigned to a lexical NP in external argument position.

If we put these two assumptions together, we then expect the optionality of Case absorption to be a parametrized property of object clitics also.

Now, if we assume, as does Borer (1984), that a dummy Case-marker is a Case-assigner then it is not possible for an object clitic not to absorb Case since, if this happened, there would be no need for a dummy Case-marker. Furthermore, it would be suspect in terms of learnability to make a distinction between Case-absorption by an object clitic and Case-absorption by a subject clitic at a cross-linguistic level. Finally, recall that PN French exhibits both object doubling and subject doubling, which makes a distinction between subject clitics and object clitics as to how they absorb Case look rather strange since this distinction would exist even in a given particular grammar.

For these reasons, I assume here that object clitics can also absorb Case optionally and that this is a parametrized option. The parameter in (43) above can thus be generalized as in (73).

(73)a. Unmarked setting:
A clitic absorbs Case.
b. Marked setting:
A clitic need not absorb Case.

With respect to learnability, it remains that the language-learner will have to establish the validity of (73) for each clitic paradigm (subject, object of V, object of P, etc.). In other words, the availability of doubling for one set of clitics does not entail the availability of doubling for all sets of clitics.

This move causes a problem for object doubling which, we now suppose, functions as subject doubling. We predict that dummy Case-markers need not be used in object-doubling constructions as is the case in subject-doubling constructions. This is an incorrect prediction, since object doubling seems to require dummy Case-markers as shown in (56)-(64) and expressed by Kayne's generalization.

One obvious solution to this problem is to consider dummy Case-markers as Case-transmitters. Jaeggli (1986) adopts such a view of the preposition *a* in the River Plate Spanish examples of object doubling, as in (56). In this view, *a* is a kind of "Case bridge" which allows the Case feature of a verb to be assigned to the NP object. Such an analysis predicts that the Case assigned to a clitic-doubled direct object is accusative and not the Case normally assigned by the preposition *a*, namely dative. Jaeggli argues that this prediction is borne out and evidence for this comes from the use of *éllo*. *Ello* is a third person neuter full pronominal form which can only occur in the dative Case. As predicted, *éllo* cannot appear in object-doubling constructions (74a) or in subject position (74b) since it is accusative in the first and nominative in the second. But it appears in regular dative assignment environments as in (74c). The sentences are from Jaeggli (1986).

(74)a. *Yo lo_i entiendo a $éllo_i$.
'I understand it.'
b. *Ello es evidente/divertido.
'It is obvious/fun.'
c. Hay que restarle a éllo lo que me acabas de dar.
'It is necessary to substract from it what you have just given me.'

If the preposition *a* in (74a) assigned dative Case to the clitic-doubled NP, then there would be no reason for the ungrammaticality of the sentence. This is thus a convincing argument from Jaeggli (1986) in favor of analyzing dummy Case-markers (at least in RP Spanish) as Case-transmitters. It is worth noting here that adopting a parameter like the one in (73) mitigates against viewing clitics as spell-outs of Case features since this parameter gives this spelling-out an undesirable optional status. This is undesirable because, under such a view, the clitic is in fact the Case itself realized morphologically on a head. Making it optional would amount to saying that the clitic sometimes is the Case and sometimes is not, which is counterintuitive.

We are left with the problem raised by the obligatory presence of a dummy Case-marker in object doubling constructions.

The real problem lies in accounting for the difference between subject doubling and object doubling with respect to dummy Case-markers.

It is proposed in Levin and Massam (1985) and Massam (1985) that two conditions must be fulfilled for an element to assign Case. The element must have both a Case feature [+C] and a Case-assigning feature [+CA]. This is the "normal" pattern when, say, a verb assigns Case to an object NP it governs. Now, let us define the term "absorption" and the associated parameter in (73) in a more precise way and suggest that a clitic always absorbs the [+CA] feature of the lexical head, but only absorbs the [+C] feature optionally as stated in (73). In languages with no doubling both [+C] and [+CA] are absorbed by the clitic, thereby making it impossible for the verb to assign Case to an object even though it governs it. In languages with object doubling, only [+C] can be freed from absorption by the clitic. Now, in order for this Case to be assigned there needs to be a [+CA] feature in the construction and that is what the role of the dummy Case-marker is. A dummy case-marker thus provides the [+CA] feature necessary for [+C] to be assigned to a doubled NP.[14] This can be extended to the construct state in (70) and to objects of prepositions in (71).[15] The structures in (69)-(71) are repeated in (75)-(77) in the light of the preceding discussion. DCM stands for "dummy Case-Marker."

(75)

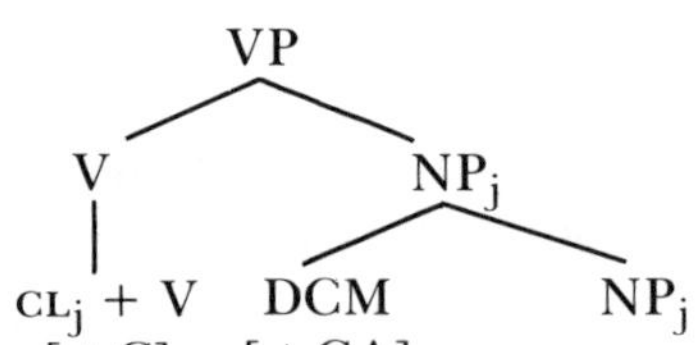

(76)

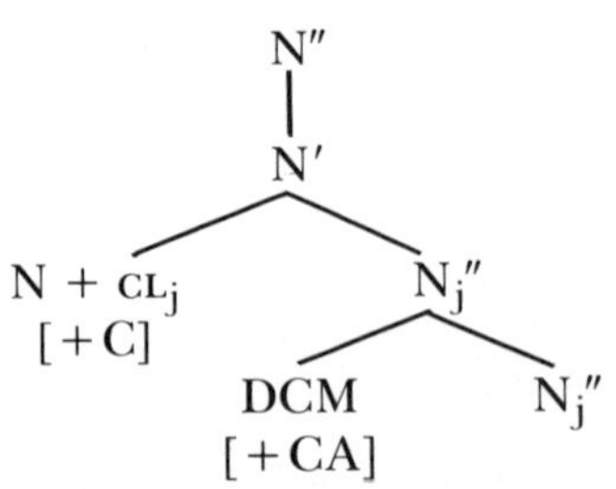

(77)

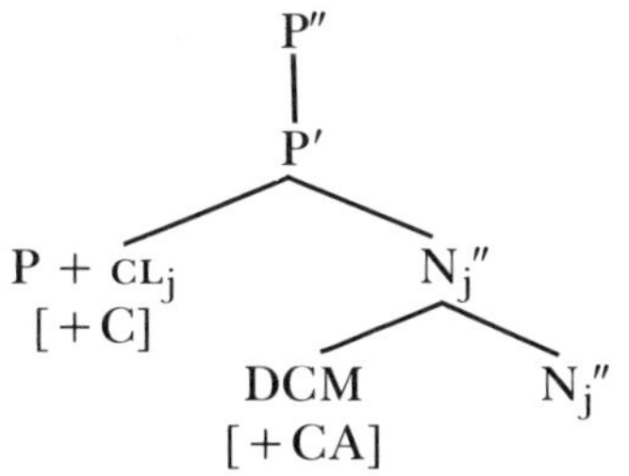

In the sentences where no clitic absorbs the [+CA] feature of the lexical head then Case assignment follows the normal path. This is illustrated in (78)-(80).

(78)

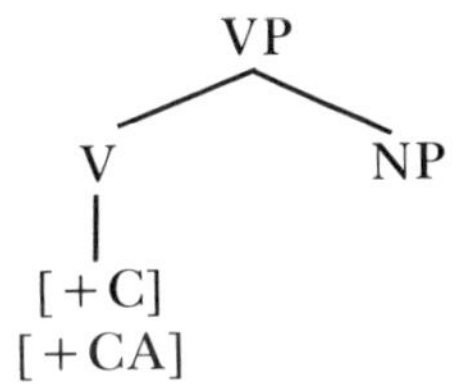

(79)

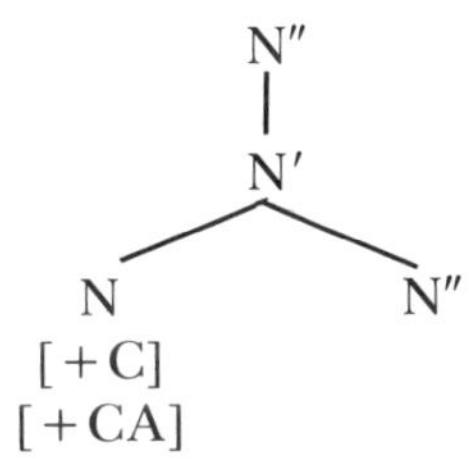

(80)

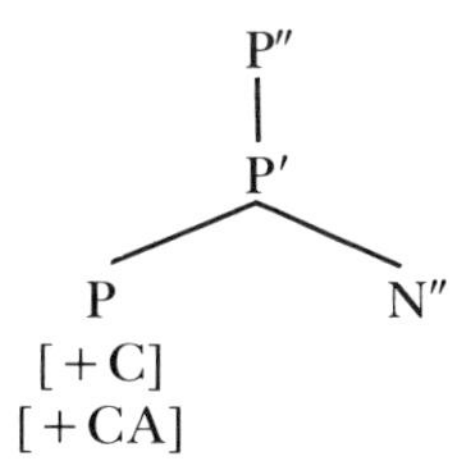

Now, with respect to the non-availability of dummy Case-markers in subject-doubling constructions, I would like to suggest two plausible and rather different accounts. The first one consists of modifying the process of nominative Case-assignment in the following way. I have already assumed in 2.4 that TNS is the nominative Case-assigner. Suppose that we qualify this by assuming that TNS provides the [+CA] feature and that AGR provides the [+C] feature. In other words, AGR carries nominative case which is assigned by TNS.

This has the following consequences. In a standard construction as in (81) with a lexical NP subject and no subject clitic, the Case of AGR is assigned to the NP by TNS.

(81)

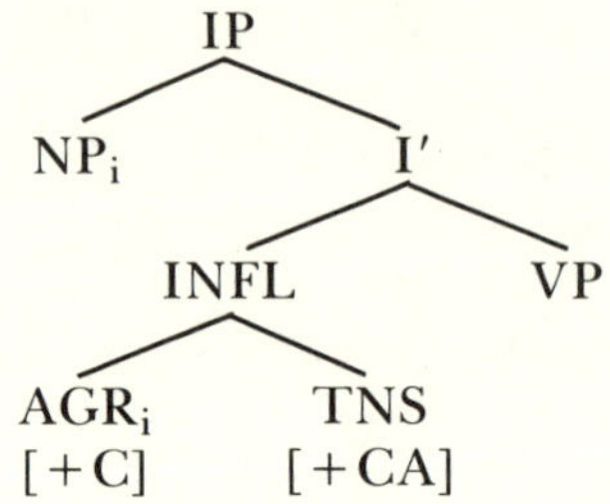

In a construction involving a subject clitic and no lexical NP as in (82), the clitic absorbs [+C] but cannot absorb [+CA] since the clitic is not part of the TNS node.

(82)

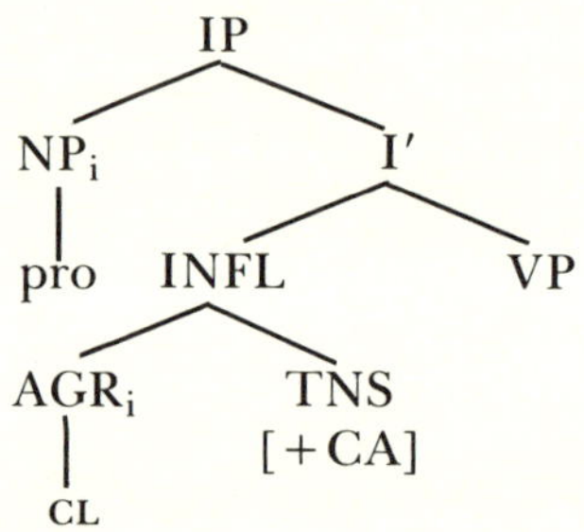

In (82), the [NP.IP] position can only be filled by pro for the familiar reasons and because pro does not need Case.

In languages with subject doubling, i.e. where the parameter in (73) is operative for subject clitics, the clitic does not absorb the [+C] feature of AGR and Case can be assigned to some lexical material in [NP,IP]. This is illustrated in (83).

(83)

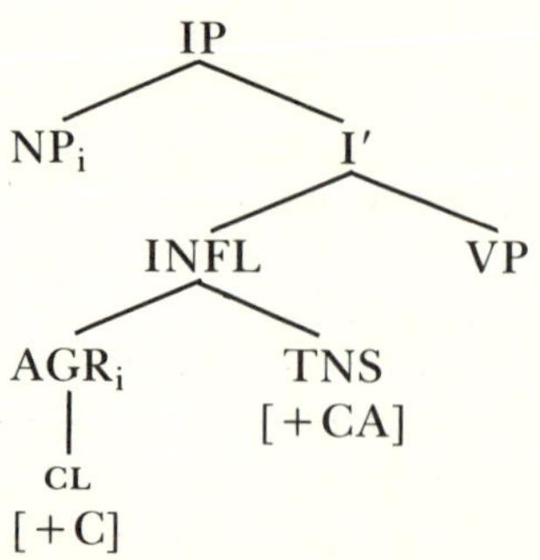

In other words, a subject clitic (optionally) absorbs [+C] in AGR but cannot absorb [+CA] in TNS, and Case can potentially be assigned to the subject position as long as a [+C] feature is available, namely, if (73) is fixed positively; i.e. if the clitic does not absorb the [+C] feature.[16] This is the first option.

The second option is less complicated but more stipulative. Let us assume that dummy Case-markers, like prepositions, can assign or transmit Case only if they are governed by a lexical head. This is the case, as we have seen, in the structures (69)-(71) in which V,N, and P govern the position where the dummy Case-markers are generated. Now, in the subject-doubling structure in (72), the situation is quite different. Since a preposition inserted in front of [NP,IP] would not appear in a governed position, it would not assign Case to the NP. In other words, the appeal to a dummy Case-marker is dictated by structural conditions and Case-assignment to the subject in subject-doubling constructions must be accounted for in a different way. Even under a conception of dummy Case-markers as Case-transmitters, note that there are no "postpositions" but only prepositions in the languages under study here which exhibit subject doubling. There is therefore no possibility for a preposition in (84) to transmit the Case released by the clitic.

(84)

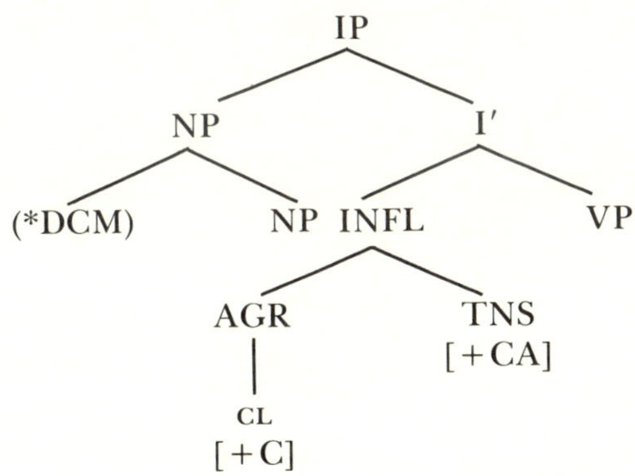

The only other means for a doubled NP to get Case would be through the regular process of nominative Case-assignment. In languages where (73) is set positively, a subject NP receives Case in the same way whether a clitic is present or not. In a language where (73) is set negatively, no case can be assigned to [NP,IP] when a clitic is present as in (85) since the clitic absorbs the [+C] feature of AGR.

(85)

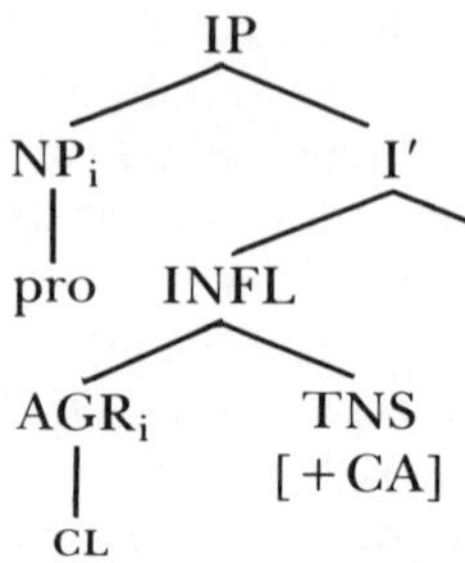

This hypothesis is not unproblematic and brings us back to the initial question of determining why dummy Case-markers are *necessary* in object-doubling constructions but not in subject-doubling constructions. If the account above is available for subject clitics, why is it not available for object clitics? In other words, why can the regular process of Case-assignment not apply in, say, (69)? Such a question finds no answer in the second account.

I thus leave the question open for the moment pending further research that would allow one to make a choice between the two options. For example, if we were to find a language with a sentence structure similar to the one in (85) but with postpositions instead of prepositions, and if this language had object doubling and subject doubling, then the second account would predict that dummy Case-markers would be used in subject doubling but not in object doubling as illustrated in (86).

(86)

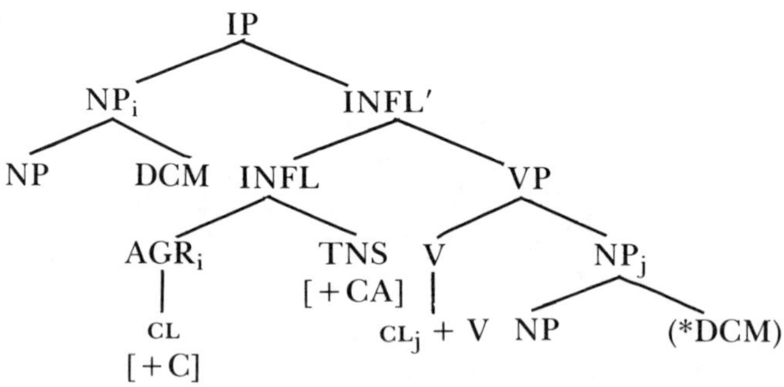

To conclude this section, it is assumed that subject clitics and object clitics do not have intrinsic differences and that the distinction between the structures in (69)-(71) where dummy Case-markers are obligatory and the one in (72) where dummy Case-markers are never used, must follow from independently required structural subject/object asymmetries. I have presented two options which allow us to maintain a single parameter, (73), for both subject clitics and object clitics.[17]

3.3. EXTRACTIONS

We now have a basic parametric account of doubling phenomena which encompasses the different instances of the phenomenon illustrated in (75), (76), (77), and (83). We are left with one major source of variation among the languages and dialects studied here, namely extraction possibilities out of doubling constructions. This is an area that has received much attention in the literature; cf. Jaeggli (1982;1984;1986), Borer (1984), and Roberge (1988) among others.

At first, when only the particular examples of object doubling in RP Spanish were examined, it was assumed that extraction out of doubling constructions were to be barred. Later on, with the introduction of more data from more languages, the situation proved more intricate than previously assumed, because other languages allowed extraction from doubling constructions similar to the RP Spanish ones.

The aim of this section is to introduce data from subject doubling and to explore the behavior of clitic-doubled NPs with respect to extraction possibilities. These data are compared with the ones studied in previous accounts. The ultimate goal is to provide a unified account of the facts. A close examination of these facts will reveal that such an account can provide a characterization of the type of lexical NPs that can be clitic doubled in various languages.

In the first part of this section, the data from seven Romance languages/dialects are presented and the different properties to be accounted for are summarized in 3.3.2. I then proceed with a review of some of the previous accounts and argue that the (im)possibility of extracting out of doubling constructions must be a parametrized property. In the final section, an account in terms of c-chain is proposed.

3.3.1. The Facts

3.3.1.1. River Plate Spanish. Jaeggli (1982) points out that a clitic-doubled object NP cannot be extracted by *wh*-movement in River Plate Spanish as shown in (87).

(87)a. Lo_i vimos a Guille_i.
CL saw-1P to
'We saw Guille.'

b. *A quién$_i$ lo$_i$ vimos t_i?
to whom CL saw-1P
'Whom did we see?'
c. A quién$_i$ vimos t_i?

Furthermore, extractions taking place at Logical Form (LF) are also barred. The quantified NP objects in the ungrammatical sentences in (88) undergo Quantifier Raising (QR) at LF.[18]

(88)a. *Las$_i$ ví a todas las chicas$_i$.
CL saw-1S to all the girls
'I saw all the girls.'
b. *Las$_i$ encontré a algunas mujeres$_i$.
CL met-1S to some women
'I met some women.'
c. *Lo$_i$ ví a un chico$_i$.
CL saw-1S to a boy
'I saw a boy.'

3.3.1.2. Colloquial French. Subject doubling in C.Fr. obeys the same type of restriction as object doubling in RP Spanish, i.e. extraction from subject position is impossible when there is a coreferent subject clitic in the same clause. This is illustrated in (89).

(89)a. *Qui$_i$ il$_i$ est allé?
who CL is gone
'Who went?'
b. *[$_{CP}$ *wh*$_i$ [$_{IP}$ t_i CL$_i$+V . . .]]
c. Qui est allé?
d. [$_{CP}$ *wh*$_i$ [$_{IP}$ t_i V . . .]]

There is also, parallel to the RP Spanish example in (88c), a definiteness requirement on the lexical NP associated with a subject clitic.

(90)a. La fille elle va chanter.
the girl CL go-3S to + sing
'The girl will sing.'
b. *Une fille elle va chanter.
a girl CL go-3S to + sing.
'A girl will sing.'
c. Une fille va chanter.

The string in (91b) must therefore be excluded.

(91)a. $[_{IP}$ NP_i CL_i + V . . .] = (90a)
[+def]
b. *$[_{IP}$ NP_i CL_i + V . . .] = (90b)
[−def]

If it is true that indefinite NPs are quantified expressions and, as such, must undergo QR at LF then, in (91b), the NP is extracted for scope interpretation requirements. The string is therefore:

(92) *NP_i $[_{IP}$ t_i CL_i + V . . .]

As expected, regular quantified NPs are also barred from subject-doubling constructions.[19]

(93)a. *Quelques $gars_i$ ils_i ont trop
a + few guys CL have-3P too + much
bu hier.
drunk yesterday
'A few guys drank too much yesterday.'
b. *$Personne_i$ il_i aime travailler.
nobody CL like-3S to + work
'Nobody likes working.'
c. *Tous les $hommes_i$ ils_i aiment manger.
all the men CL like-3P to + eat
'All men like eating.'
d. *N'importe qui_i il_i peut faire ça.
anybody CL can make this
'Anybody can do that.'

The syntactic representation for the sentences in (93) is also the one in (92), which is itself very similar to the one in (89b). What I want to suggest here and for the rest of this section is that (90b) and (93) are excluded for the same reason (89a) is. As pointed out in Jaeggli (1982), it is generally impossible to extract a clitic-doubled NP. Furthermore, the fact that QR as well as *wh*-movement is excluded suggests that the constraint responsible for blocking movement holds at S-structure as well as LF, at least in Colloquial French and River Plate Spanish.

I will continue to assume that the definiteness effect in (88c) and (90b) is in fact a "no-quantifier effect".

3.3.1.3. Pied Noir French. As stated earlier (3.1.3.2), Pied Noir French exhibits a subject-doubling property very similar to the C.Fr.

one in that they are both optional. But PN Fr. differs from C.Fr. in that quantified NPs can be clitic-doubled in subject position. As in C.Fr., *wh* words or phrases are excluded. The relevant data is found in (94).

(94)a. L' homme$_i$ il$_i$ vient.
the man CL come-3s
'The man comes.'

b. Un homme$_i$ il$_i$ vient.
a
'A man comes.'

c. Plusieurs femmes$_i$ elles$_i$ sont venues.
many women CL have-3P come
'Many women came.'

d. Personne il sait qui c' est leur mère.
nobody he know-3s who it be-3s their mother
'Nobody knows who their mother is.'

e. *Qui$_i$ il$_i$ vient?

The preliminary conclusion in light of these facts is that the constraint against extraction out of subject-doubling construction holds only at S-structure in PN Fr. since only *wh*-movement (94e), but not QR (94b,c,d), is blocked.

Now, a very interesting aspect of PN Fr. syntax referred to in footnote 8 becomes extremely relevant here. Recall that both subject doubling and object doubling are attested in this dialect. We thus have a direct way of testing our hypothesis that subject doubling and object doubling are two different realizations of the same phenomenon and are to be accounted for in the same way. According to this hypothesis, the constraint on extraction from subject doubling in PN Fr. should apply in exactly the same way it does in object-doubling constructions. More precisely, it should be possible to find quantified NPs in object-doubling constructions where *wh* words would be disallowed. That this is true is illustrated in (95) and (96).[20]

(95)a. Jean l$_i$' a vu à plusieurs personnes$_i$.
CL have-3s seen to many people
'Jean saw many people.'

b. Leurs complices les$_i$ ont accusés à
their accomplices CL have-3P accused to
tous les voleurs$_i$.
all the thieves
'Their accomplices have accused all the thieves.'

c. Il les enterre à tous au cent mètres.
he CL bury-3s to all at + the hundred metres
'He beats them all at the one hundred metres.'

d. *À qui_i que tu l_i' as vu t_i?
to whom that you CL have-2s seen
'Who did you see?'

e. *À qui_i que tu le_i frappes t_i?
to whom that you CL beat-2s
'Who are you beating?'

Naturally, non-quantified NPs are also allowed to be clitic-doubled. Compare (96) to (95a,b,c).

(96)a. Marie l_i' a vu à $Jean_i$.
CL have-3s seen to
'Marie saw Jean.'

b. Tu le_i frappes au petit $gars_i$.
you CL beat-2s to + the little guy
'You beat the little guy.'

Furthermore, it is interesting to note that, contrary to the case in RP Spanish, a clitic-doubled NP can be focused in PN Fr. as illustrated by the discourse in (97).[21]

(97) A: Savais-tu que son complice a accusé Pierre?
'Did you know that his accomplice accused Pierre?'
B: Non! Son complice l_i'a accusé à $JEAN_i$.
'No! His accomplice accused JEAN.'

If a rule of focus interpretation raises the NP at LF then the grammaticality of the construction in (97b) is expected given that quantified NPs (also raised at LF) are allowed in the same context, as shown in (95a,b,c). See Chomsky (1976), Culicover and Rochemont (1983) and Rochemont (1986) for detailed analyses of the representation of focus.

My preliminary conclusion thus holds: the constraint against extraction out of doubling applies at S-structure but not at LF in PN Fr.

3.3.1.4. Trentino. The behavior of subject clitics and lexical NP subjects was presented in 3.1.1. Recall that, in this dialect, subject clitics are obligatory either in isolation or when a lexical preverbal NP is present. They are excluded in the free inversion construction.

The sentences in (98) illustrate that there is no "no-quantifier requirement" in subject doubling in Trentino, i.e. quantified NPs and indefinite NPs can freely co-occur with a coreferent subject clitic.

(98)a.	Qualche	putel$_i$	l$_i$'	é	vegnú.
	some	boys	CL	be-3P	come
	'Some boys came.'				
b.	Una	putela$_i$	l$_i$'	ei	vegnuda.
	a	girl	CL	be-3S	come
	'A girl came.'				

It is thus plausible to assume that extraction of a preverbal clitic-doubled NP subject is possible in this N.It. dialect. But the facts are not so simple and a qualification is in order here. In effect, *wh*-movement cannot take place in this construction as shown in (99) to (102) where a subject clitic cannot surface after subject extraction. The clitics are underlined.

(99)a.	Quante	putele	é	na	via?
	how + many	girls	be-3P	gone	away
	'How many girls have left?'				
b.	*Quante putele$_i$ é<u>le</u>$_i$ nade via?				
c.	*Quante putele$_i$ <u>le</u>$_i$ é nade via?				

(100)a.	Chi	ha	magna?
	who	have-3S	eaten
	'Who ate?'		
b.	*Chi$_i$ ha<u>lo</u>$_i$ magna?		
c.	*Chi$_i$ <u>el</u>$_i$ ha magna?		

(101)a.	Quante	putele	penset	che	sìa	vegnú?
	how + many	girls	think-2S	that	be-3P	come
	'How many girls do you think came?'					
b.	*Quante putele$_i$ penset che <u>le</u>$_i$ sìa vegnude?					

(102)a.	Chi	penset	che	magna?
	who	think-2S	that	eat-3S
	'Who do you think is eating?'			
b.	*Chi$_i$ penset che <u>el</u>$_i$ magna?			

It must thus be assumed that whatever constraint is violated in (99)-(102) crucially holds at S-structure but not at LF in order to allow for (98) where QR applies.

3.3.1.5. Fiorentino. In this other N.It. dialect, the facts are basically the same as in Trentino, i.e. subject clitics are obligatory either in isolation or when a lexical preverbal NP subject is present as shown in (103).

(103) Qualche ragazzo$_i$ gl$_i$' é venuto.
some boy CL be-3S come
'Some boy came.'

One of the major differences between Fiorentino and Trentino concerns the presence of a preverbal impersonal clitic in free inversion constructions. Contrast (104) with (105), where no impersonal clitic appears in preverbal position.[22]

(104) Fiorentino
a. *Gl'* é venuto una ragazza.
CL(IMP) be-3S come a girl
'A girl came.'
b. *Gl'* ha telefonato una ragazza.
CL have-3S telephoned a girl
'A girl telephoned.'
c. *Gl'* ha portato la torta una ragazza.
CL have-3S brought the cake a girl
'A girl brought the cake.'

(105) Trentino
E' vegnú 'na putela.
have-3S come a girl
'A girl came.'

As far as *wh*-movement is concerned, the impersonal clitic surfaces in preverbal position when it applies to a subject as in (106).

(106)a. Una ragazza$_i$ *l*$_i$' é venuta.
a girl CL be-3S come-F
'A girl came.'
b. *Gl'* é venuto una ragazza.
CL(IMP) be-3S come-M a girl
'A girl came.'
c. **L*'é venuta una ragazza.
d. Chi *gl*'é venuto?
e. *Chi é venuta?
f. *Chi *l*'é venuta?

The sentences in (106a,b,c) show that only the impersonal clitic can be used with a postverbal subject. In (106d), the subject is extracted and the impersonal clitic is present; this is grammatical. The ungrammatical cases in (106e,f) are blocked because the subject is extracted from preverbal position either without a clitic (e) or with a preverbal referential clitic (f). As pointed out in Jaeggli (1984), this is evidence that subject extraction takes place from postverbal position in Fiorentino. I return to subject extraction in Trentino and Fiorentino in 3.3.4.

Thus extraction can take place at LF out of subject-doubling construction in Fiorentino since (103) and (106a) are grammatical, but *wh*-movement is not allowed, as shown by the ungrammatical sentences in (106e,f).

The situation is therefore the same as it is in Trentino with respect to the constraint blocking extractions from subject doubling, i.e. it applies at S-structure but not at LF.

3.3.1.6. Romanian. The Romanian data is considerably more complex but it is possible to concentrate on the relevant well-documented generalizations. The Romanian examples are taken from Borer (1984) and Steriade (1980); cf. also Farkas (1978) and Dobrovie-Sorin (1985) for more on clitic-doubling in Romanian.

Clitic-doubling of a direct object in Romanian is restricted in the familiar way to [+human/pronominal], [+specific/definite] NPs. In the sentences below OM stands for "object marker" and the clitics are underlined.

(107)a. Am väzut Popescu.
have-1s seen
'I have seen Popescu.'

b. *L* -am văzut.
CL-have-1s seen
'I have seen him.'

c. *L_i*-am văzut pe Popescu$_i$.
CL-have-1s seen OM
'I have seen Popescu.'

d. **L_i*-am văzut Popescu$_i$.

e. *Am văzut pe Popescu.

(108)a. Am văzut un bucătar.
have-1s seen a cook
'I have seen a cook.'

b. *L_i*-am văzut.

c. *L_i-am văzut pe un bucătar$_i$.
d. *L_i-am văzut un bucătar$_i$.
e. *Am văzut pe un bucătar.

(109)a. Am văzut pe alticineva.
have-1s seen OM somebody else
'I have seen somebody else.'
b. *L_i-am văzut pe alticineva$_i$.

The impossibility of object doubling in (108c) and (109b) is to be attributed to the nature of the doubled NP, namely [−specific, −definite, +human, −pronominal] in (108) and [−specific, −definite, +pronominal] in (109). On the other hand, a [+specific, +human, +pronominal] NP such as the one in (107) can be clitic-doubled as in (107c).

Note that according to what we have seen to this point, it is hypothesized that a [+specific, −definite] NP that is also [+human] should be allowed in object-doubling constructions in Romanian. This is true. Therefore, (108c) is grammatical if the indefinite NP *un bucătar* is [+specific].

The sentence in (110) is also grammatical because [$_{NP}$ *o fată de la noi din sat*], although it is indefinite is also possibly specific.

(110) O_i caut pe [o fată de la noi
CL be + looking + for-1s OM a girl of the our
din sat]$_i$.
village
'I am looking for a girl from our village.'

This situation is predicted here under the hypothesis that a specific indefinite NP does not undergo QR at LF as regular quantified NPs (109) and non-specific indefinite NPs (108) do. It is thus natural to assume that the constraint against extractions out of doubling constructions in Romanian holds at LF.

The data involving *wh*-extractions is more problematic. Consider the relative and interrogative clauses in (111) to (113).

(111)a. Casa pe care$_i$ credeai că am
the + house OM which thought-2s that have-1s
văzut-o$_i$ . . .
seen -CL
'The house which you thought that I have seen . . .'

b. Pe care$_i$ credeai că am văzut-o$_i$?
OM which thought-2s that have-1s seen CL
'Which one did you think that I have seen?'

(112)a. Pe cine credeai că am văzut?
OM who thought-2s that have-1s seen
'Whom did you think that I have seen?'
b. *Pe cine$_i$ credeai că am văzut-o$_i$?

(113)a. Ce credeai că am văzut?
what thought-2s that have-1s seen
'What did you think that I have seen?'
b. *Ce$_i$ credeai că am văzut-o$_i$?

At first sight, it seems as if *wh*-extractions out of object doubling are possible in Romanian since this is presumably what happens in (111). The appearance of a clitic in (112b) and (113b) is blocked by the constraint above on the nature of the NP object that can be doubled since the extracted NP is non-specific; cf. Steriade (1980). On the other hand, object doubling is possible in (111) because the doubled (extracted) NP is specific.

A preliminary conclusion would thus be that only *wh*-extractions are allowed out of object-doubling constructions in Romanian.[23] As we have seen in (108c) and (109b) above, LF extractions are barred.

However, it is not clear that *wh*-extractions out of object-doubling constructions are allowed in Romanian. Dobrovie-Sorin (1985, 9) argues that "Romanian, just like River Plate Spanish, does not allow *wh*-movement of doubled direct object; in other words, *wh*-movement structures exclude clitics corresponding to the moved constituent." I will not describe here the detail of her arguments but will simply review the assumptions underlying them.

Dobrovie-Sorin examines the Romanian clauses below where *e* stands for an empty category.

(114) *[pe care]$_i$ ai văzut e_i
OM which have-2s seen

(115) [pe cine]$_i$ ai văzut e_i
OM who

(116) [pe care]$_i$ l$_i$-ai văzut e_i
CL

(117) *[pe cine]$_i$ l$_i$-ai văzut e_i

These clauses illustrate the structural possibilities corresponding to interrogative clauses; they naturally extend to relative clauses.[24] She argues that the correct LF representation for *cine*-sentences and *care*-sentences are (118) and (119) respectively.

(118) *cine*-sentence
for which x, you saw x

(119) *care*-sentence
$[_{NP}$ for which x, (x=N') (or x=PRO)$]_i$ you saw e_i

Pe cine is thus a quantifier binding a variable in IP. Of *pe care* she says (p.7): "*Care* is a quantifier, but it does not bind a variable in S (=IP). We claim that in Romanian *care* constructions the scope of the quantifier is limited to the NP in COMP; the quantifying domain of *care* is not the class of the elements that can fill the argument position (the one occupied by e_i) in S, but PRO itself, more precisely the class of elements that satisfy the referential properties defined by the antecedent of PRO: the class of girls, boys, etc." Suppose that this is true. We can now return to the clauses in (114)-(117). In (114) the variable is not $\bar{A}$-bound since *pe care* cannot be the binder. Alternatively e_i could be pro but this occurrence of pro would not be licensed according to our hypotheses about the nature of the relationship between pro and a clitic; the sentence is thus ungrammatical. *Pe cine* in (115) is a proper $\bar{A}$-binder for e_i; nothing is violated and the sentence is grammatical. In (116) e_i is not a variable as in (114), it is thus presumably pro which is licensed by the clitic. In (117), since *pe cine* binds e_i, this empty category is a variable but it is also related to a clitic, which is unacceptable given the constraint against extraction out of doubling constructions that will be introduced in 3.3.4.

A more plausible conclusion concerning the Romanian data is therefore that LF extractions as well as *wh*-extractions are blocked in doubling constructions. This conclusion will be maintained throughout the remainder of this section although I am aware of the fact that Dobrovie-Sorin's account is based on the semantic interpretation of the relationship between *care* and the empty category in internal argument position.

3.3.1.7. Standard French. The situation in Standard French is the following: there is no free inversion (120), subject doubling is barred (121) and there can be no *wh*-movement of an NP if a corresponding clitic is present in the structure (122). Note that the latter property could be interpreted as a consequence of the unavailability of doubling in this dialect.[25]

(120)a. *Est venu Pierre.
is come
'Pierre has come.'
b. Pierre est venu.

(121)a. *Pierre$_i$ il$_i$ est venu.
CL
b. *Les trois filles$_i$ elles$_i$ ont bu.
the three girls CL have-3s drunk
'The three girls drank.'

(122)a. *Qui$_i$ il$_i$ est venu?
who CL be-3s come
'Who has come?'
b. Qui est venu?
c. *Combien de filles$_i$ dis-tu qu' elles$_i$
how + many of girls say you that CL
ont bu du thé?
have-3s drank of tea
'How many girls did you say drank tea?'

3.3.2. Summary

We can now establish some basic ways in which the languages/dialects under study here can differ. The different properties are the ones in (123).

(123)a. +/− subject doubling
b. +/− object doubling
c. +/− *wh*-extraction of a clitic-doubled NP
d. +/− LF extraction of a clitic-doubled NP

These properties are distributed in the following way among the languages, where "+" means that the property exists, "−" means that the property does not exist, "ϕ" means that the property is not relevant either because the language does not have the relevant set of clitics or because there is no doubling and hence no possibility (or impossibility) of extracting out of doubling; but see footnote 25.

(124) *Subject Clitics*

property / language	doubling	*wh*-extraction	LF extraction
RP Spanish	ϕ	ϕ	ϕ
C. French	+	–	–
PN French	+	–	+
Trentino	+	–	+
Fiorentino	+	–	+
Romanian	ϕ	ϕ	ϕ
St. French	–	ϕ	ϕ

(125) *Object Clitics*

property / language	doubling	*wh*-extraction	LF extraction
RP Spanish	+	–	–
C. French	–	ϕ	ϕ
PN French	+	–	+
Trentino	–	ϕ	ϕ
Fiorentino	–	ϕ	ϕ
Romanian	+	–	–
St. French	–	ϕ	ϕ

Any account of the relationship between clitics and argument positions will have to leave open:

(126) A. the option of having doubling in the first place;
B. the option of extracting out of doubling in syntax;
C. the option of extracting out of doubling in LF.

Such an account is suggested in 3.3.4 but before introducing it, I want to review some previous accounts of some of the facts mentioned above.

3.3.3. Previous Accounts

To my knowledge none of the analyses already available can handle the three options in (126) at the same time. I review three of them here.[26]

The first analysis is that of Jeaggli (1982). In this analysis, proposed to account for the impossibility of extracting a clitic-doubled NP in object position in River Plate Spanish, it is assumed that an object clitic affixed to V absorbs both the government and the Case-assigning properties of V. The empty category in the ungoverned object position is PRO.

This analysis makes strong predictions with respect to extraction. In effect, the Empty Category Principle (ECP) stated in (127) will rule out any occurrence of a trace in a clitic-doubled object position since this trace would not be properly governed.

(127) [$_{\alpha}e$] must be properly governed.

(Chomsky 1981, 250)[27]

Note that with the ECP applying at LF, extractions in syntax as well as in LF are blocked in the same way since a trace resulting from an application of Move α in syntax is still a trace at the LF level. It is therefore impossible to account for the options in (126b,c).

The second analysis is that of Borer (1984). For Borer, the definition of proper government includes a condition on Case-agreement between the governor and the governee. She argues further that a clitic properly governs the argument position it is related to if, among other things, the clitic and this position bear the same Case. Since a clitic is, in her account, a spell-out of Case features on a head, it follows that the argument position receives Case from the so-called dummy Case-marker. Consider again the River Plate Spanish and Romanian sentences in (87b) and (111b) repeated in (128).

(128)a.	*A	quién$_i$	lo$_i$	vimos	t_i?		
	DAT	whom	CL(ACC)	saw-1P	DAT		
b.	Pe	care$_i$	credeai	că	am	văzut-o$_i$	t_i?
	ACC	which	thought-2S	that	have	seen -CL(ACC)	ACC

In (128a) the preposition *a* assigns dative Case to the object *quién*. After *wh*-extraction, the trace is also marked dative. The object clitic *lo* is accusative. There is therefore Case mismatch between the governor and the governee and the ECP rules out the sentence since the trace is not properly governed.

In (128b) the situation is quite different according to Borer. The preposition *pe* assigns accusative Case, the trace is thus marked ACC but the object clitic *o* is also accusative. The Case matching requirement being met, the clitic can properly govern the trace, making extraction possible.

This type of analysis relies heavily on the use of dummy Case-markers and abstract Case to characterize language variation. This causes a problem in subject-doubling constructions where this particular type of "rescuing device" is not appealed to; cf. 3.2.2. In this particular construction, there is no reason to assume that the clitic-doubled NP and the clitic itself do not bear the same (nominative) Case. How then can the illicit extractions either in syntax or at LF be explained?

Moreover, Jaeggli (1986) points out that there is no evidence suggesting that the trace in (128a) is actually dative. In fact, certain facts concerning the use of the third person neuter full pronoun lead him to conclude that this trace is accusative; cf. 3.2.2. If this is true, then there is no distinction with respect to Case-marking between RP Spanish and Romanian in (128).

Finally, the interpretation of the Romanian data adopted here is that of Dobrovie-Sorin (1985) who argues that *pe care* in (128b) does not bind a variable (t_i); cf. 3.3.1.5. The contrast in (128) is therefore merely apparent and there is no difference between RP Spanish and Romanian with regard to extractions out of doubling constructions as illustrated in (125); but see footnotes 25 and 31 for some problematic cases in Modern Hebrew.

The last analysis is that of Jaeggli (1984). This analysis constitutes one of the most recent attempts at explaining the facts about extraction and inversion. Jaeggli presents an account of the impossibility of certain syntactic extractions in which no appeal is made to the ECP. His approach is motivated by the fact that *wh*-movement from preverbal position to the nearest COMP is blocked in Italian. Extraction must take place from the postverbal position. Consider the contrast in (129).

(129)a. Quante ne sono cadute?
how + many of + them be-3P fallen
'How many of them have fallen?'
b. *Quante sono cadute?

Italian *ne* is a clitic affixed to V which is related to an empty category in an NP internal to VP, i.e. not Chomsky-adjoined to VP. *Ne*-cliticization is thus possible in (130b) with the structure in (130c) but not in (131b) the structure of which is (131c).

(130)a. Arrivano molti studenti.
arrived many students
'Many students arrived.'

b. Ne arrivano molti.
c. $[_{VP}$ arrivano $[_{NP}$ molti studenti]]

(131)a. Telefonano molti studenti.
telephoned many students
'Many students telephoned.'
b. *Ne telefonano molti.
c. $[_{VP}$ $[_{VP}$ telefonano] $[_{NP}$ molti studenti]]

The presence of *ne* in (129a) is an indication that the subject is in postverbal position at D-structure and extraction of the equivalent of *molti* in (130) and (131) is allowed. If, as in (129b), the subject is in preverbal position (there is no *ne*-cliticization) then extraction is barred; see Belletti and Rizzi (1981) for more on *ne*-cliticization.

The point here is that if it is assumed that postverbal extraction is forced by the ECP, then there is no reason why it should apply in (129). The S-structure representation for (129b) is in (132).

(132) $[_{CP}$ wh_i $[_{IP}$ t_i INFL VP]]

The trace is properly governed by the *wh*-phrase in COMP, yet the structure is rejected. This is what Jaeggli (1984, 139) attempts to explain.

To do so, Jaeggli adopts a contextual definition of empty categories according to which the empty category in (132) is actually pro in Italian since rich agreement governs it. We thus have (133) instead of (132).

(133) $[_{CP}$ wh_i $[_{IP}$ pro_i INFL VP]]

Jaeggli further assumes the following:

(134) Pro cannot function as a variable if locally bound by an operator.

Hence if extraction in (129b) effectively takes place from a preverbal position, then pro is interpreted as a variable and the sentence is excluded in accordance with (134). If extraction is from a postverbal position, then the pro subject is not coindexed with the operator and the sentence is grammatical as in (129a).

Jaeggli claims that this approach can be extended to certain facts in Trentino since, as we have seen in 3.3.1.4, preverbal extraction is impossible in most cases of *wh*-movement. In this dialect, extractions of the subject take place from postverbal position.[28]

The major problem with this claim concerns the no-quantifier requirement in doubling constructions in certain languages. As shown in 3.3.1.4 and 3.3.1.5, there seems to be no such requirement in Trentino and Fiorentino. Since indefinite and quantified NPs undergo QR at LF, then there is, in fact, extraction from preverbal position, which, in Jaeggli's view of empty categories, leaves a pro locally bound by an operator; cf. also 1.3. The options in (126) are not accounted for. To account for them would require a parametrization of (134) which would apply only at S-structure or only at LF in different languages. This type of parametrization is not ideal, however, since it refers to the definition of an empty category itself, here pro. If the theory is to allow different definitions of empty categories at different levels of representation in different languages, then it loses explanatory adequacy in a particularly crucial domain.

Another potential problem for Jaeggli's analysis has to do with the Portuguese facts. In this language, it is possible to extract the subject even though there is rich subject-verb agreement but no free inversion. It can therefore not be argued that subject extraction originates from postverbal position. Since Portuguese is a null subject language similar to Italian, then, according to a definition of empty categories such as Jaeggli's, the result of subject extraction in this language is something like (133), which violates (134). Nevertheless, extractions of a preverbal subject are apparently grammatical in Portuguese.

(135) Portuguese

Que	rapazes	acreditas	que	tenham	gasto
which	children	believe-2S	that	have-3P	spent

esse	dinheiro?
that	money

'Which children do you believe spent that money?'

(Zubizaretta 1982a, 4)

To summarize, we have seen that two analyses in terms of the ECP – Jaeggli (1982) and Borer (1984) – fail to make the necessary distinction between syntactic extraction and LF extraction out of doubling constructions. Jaeggli's (1984) account, which does not rely on the ECP, still cannot handle the fact that a particular grammar can allow LF extractions while blocking syntactic extractions in the same doubling construction.

3.3.4. C-Chains

Let us continue to assume that a clitic must be related to an argument position. Borer (1984, 118) states essentially the same condition when she mentions that "the configuration [CL$_i$[e_i]] is a discontinuous pronominal element." Let us push this intuition further by assuming that a clitic and the associated argument position form some kind of a chain. I will call this particular type of chain a "c-chain" and define it as in (136).

(136) C-chain (definition)
α and β form a c-chain if:
i. α and β are coindexed;
ii. α governs β;
iii. β is locally free.

Conditions (i) and (ii) on the formation of a c-chain are the ones necessary for the licensing of an empty pronominal in the argument position associated with the clitic and are discussed extensively in chapter 2. The last condition in (iii) is new and is introduced in order to block any extraction out of the argument position (β) related to a clitic (α) since β would otherwise be locally $\bar{A}$-bound by the coindexed antecedent. The same condition rules out (overt or empty) anaphors in doubling constructions since they are locally A-bound; see 3.5 concerning reflexive clitics. Pronominals, which are locally free, and R-expressions, which are free, are not ruled out if, that is, they can somehow receive Case: cf. 3.2.2.

Note that the definition of c-chain in (136) applies not only to clitics but also to whatever element is coindexed with and governs another locally free element. This is the case in some instances of agreement markers in simple tensed sentences. There is no explicit reference to "clitics" in (136) for the reason that we would eventually want to use (136) as a diagnostic for differentiating clitics and agreement markers. Let us suppose that one is faced with data that are problematic as far as the distinction between clitics and agreement markers is concerned. In such a situation, one would want to apply tests in order to decide whether an affix on a head is a clitic or an agreement marker. These two elements are often hard to distinguish, partly because, up to a certain point, they both behave in the same way with respect to (136).

One of the ways in which (136) can be used to distinguish clitics and agreement markers is by adopting the well-formedness condition on the occurrence of clitics in (137).

(137) A clitic must be part of a c-chain.

This condition cannot be derived from the Projection Principle since only the presence of pro derives from such a principle. On the other hand, the Recoverability Principle presented in the introduction, which requires that pro and other empty categories be licensed, could be the principle that ultimately motivates the condition in (137)[29].

In the remainder of this section we will see how the condition in (137) and the definition in (136) can account for the properties presented in 3.3.1 in various languages/dialects. The level at which this condition holds will play a crucial role in my characterization of the properties in (124) and (125).

In RP Spanish as well as in Romanian there is no extraction – QR or *wh*-movement – out of object-doubling constructions. The condition in (137) therefore holds at LF and blocks the appearance of a variable associated with a clitic. The trace left by *wh*-movement or QR violates (iii) in (136) and the sentences involving them are ruled out.

The properties of C.Fr. with respect to doubling are the same as in RP Spanish except that doubling in C.Fr. is limited to the subject position. The analysis is thus the same, namely that the condition in (137) holds at LF as in RP Spanish and Romanian.

In Trentino and Fiorentino the situation is different: LF extractions out of subject doubling are allowed whereas *wh*-extractions are barred. This can be taken care of if we assume that the condition in (137) holds only at S-structure where it blocks the variable left by *wh*-movement. A variable resulting from the application of QR is not filtered out. LF extractions are thus grammatical.

Now, with respect to Jaeggli's (1984) conclusion that *wh*-movement of a subject in Trentino takes place from postverbal position, we are left with two plausible hypotheses in interpreting the data:

A. Extraction from postverbal position is forced by the fact that condition (137) holds at S-structure and would otherwise be violated.
B. Extraction takes place from postverbal position but, because of (137), a clitic cannot be present at S-structure together with the variable.

The test case that can provide us with evidence in favor of A or B is a Fiorentino interrogative clause in which the subject is questioned. Recall that, in the inverted constructions in Fiorentino, the subject appears postverbally but a neuter clitic appears preverbally. Hypothesis A predicts that such an interrogative, corresponding to (138a), would have the form in (138d), not the one in (138e). Hypothesis B, on the other hand, predicts that (138e) is the grammatical output.

(138)a. Una ragazza *l'é* venuta.
b. *Gl'é* venuto una ragazza.
c. **L'é* venuta una ragazza.
d. Chi gl'é venuto? $[_{CP}\ wh_i\ [_{IP}\ pro_j\ CL_j + V\ t_i]]$
e. Chi é venuta? $[_{CP}\ wh_i\ [_{IP}\ t_i\ V]]$

Dietter Wanner has pointed out to me that (138d) is the grammatical interrogative; (138e) is rejected. Hypothesis A is therefore the right one, which reinforces Jaeggli's conclusion. What is important here is that both hypotheses are compatible with my suggestions.

I now turn to PN Fr. in which LF extractions out of either subject-doubling or object-doubling constructions are allowed but where *wh*-extractions are barred. We see here that the situation is the same as it is in Trentino and Fiorentino with the exception that free inversion is found in the latter two dialects. The analysis is the same, i.e. (137) applies only at S-structure in PN Fr.

Note with respect to this type of parametrization that trigger experiences are available in each case. Suppose that in the unmarked case, (137) holds at LF. The language-learner then has to change this setting and this is done by exposure to sentences involving clitic doubling of a quantified NP. Since a quantified NP is allowed, (137) cannot hold at LF and will thus apply earlier in the derivation, i.e. at S-structure.

The parameter involved here would thus have the form in (139).

(139)a. Unmarked setting:
(137) holds at LF.
b. Marked setting:
(137) holds at S-structure.

Consider, finally, the case of Standard French – or any other language with clitics that does not allow clitic doubling. Needless to say, (137) holds at LF in this type of language; that is the unmarked case. The option of having doubling at all is blocked because of the conditions on Case-assignment to a doubled NP presented in 3.2.2. If a lexical NP is present together with a clitic, then this NP is not Case-marked since the clitic absorbs Case and the NP is ruled out by the Case Filter.[30]

The analysis just introduced makes at least two predictions that I now explore. They both concern "possible" languages. It must be noted, first, that the present account is designed in such a way that there should be no particular grammar allowing clitic doubling in which *wh*-extractions are allowed out of doubling constructions while quantified NPs are illicit in the same construction. Given the model of grammar adopted here, including different levels of syntactic representation, the condition in (137) will block the quantified NPs at LF. In fact, it blocks the presence of a variable associated with a clitic. If variables are blocked at LF, then the empty category left behind by *wh*-movement in the transformational component will also be excluded once the structure is fed to LF since this empty category is a variable. In other words, the possibility illustrated in (140) is excluded. Compare (140) to (124) and (125).

(140)

property / language	doubling	*wh*-extraction	LF extraction
L	+	+	−

The second prediction is that languages allowing both *wh*-extractions and LF extractions out of doubling constructions should exist. It is the representation in (141).

(141)

property / language	doubling	*wh*-extraction	LF extraction
L	+	+	+

Argentinian Spanish and Uruguayan Spanish seem to exhibit this set of properties. I will treat them as a single dialect; they are studied extensively in Hurtado (1984a;1984b;1985). Consider the sentences in (142).

(142)a. A quién$_i$ lo$_i$ acusó su complice t_i?
to whom CL accused his accomplice
'Whom did his accomplice accused?'
b. Sus complices los$_i$ acusaron a todos los estudiantes$_i$.
their accomplices CL accused to all the students
'Their accomplices accused all the students.'
(Hurtado 1984b)

In (142a) the *wh*-phrase *a quién* is extracted even though a clitic is present. The same is true of (142b) but, in this sentence, a quantified clitic-doubled NP is extracted at LF. This Spanish dialect thus patterns with the description in (141). At least three analyses can be proposed to handle this situation. Let us say that, as in the unmarked case, condition (137) holds at LF in this dialect. The only difference between it and the other languages/dialects studied above is that, in the former, condition (iii) of (136) is somewhat relaxed therefore allowing Ā-bound empty categories in its c-chains. Sentences similar to the ones in (142) can be considered to be trigger experiences for the language learner to relax condition (iii). Alternatively, it might be the case that the verbal affixes in (142) are not actual clitics but, rather, agreement markers. If this is the case, then (137) does not apply and the sentences in (142) are not problematic. Yet another possibility would be to claim that the empty category in (142a) is not a variable bound by *a quién*. In that case, this particular example would not constitute a counter-example to (136iii). The latter analysis is reminiscent of Dobrovie-Sorin's (1985) account of the Romanian data discussed in 3.3.1.6.[31]

To summarize, the four possibilities are accounted for, namely, no extraction at all, extraction in LF only, extraction in syntax and LF, and extraction in syntax only. The last one, shown in (140), is excluded under my account. The possibilities are summarized in (143) with the appropriate languages/dialects. The type of doubling (subject or object) is set aside in (143).

(143)

property / language	doubling	*wh*-extraction	LF extraction
RP Spanish C. French Romanian	+	–	–
Argentinian, Uruguayan Spanish	+	+	+
PN French Trentino Fiorentino	+	–	+
no language	+	+	–

3.4. A NOTE ON COMPLEX INVERSION

The Complex Inversion construction in Standard French is well studied in the generative literature. The reader is referred in particular to Kayne (1972) but also to Davis (1987) Jaeggli (1982), Kayne (1984), Safir (1981;1985), Sportiche (1983), and references cited there for more complete discussions. The aim of this section is to explore the way that the theory of clitics introduced in the preceding sections applies to the Complex Inversion data. I will not be concerned here with the position of the clitic with respect to the verb. The subject clitic is affixed to the verb and its suffix status is not studied here; the references above provide accounts of this fact. What we are interested in is the apparent subject-doubling property exhibited by this construction. Consider (144).

(144)a. Jean viendra -t-il?
come + FUT-3S -he
'Will Jean come?'
b. Marie a -t-elle fait cela?
have-3S -she done that
'Did Marie do that?'

In this chapter we have been concerned with the properties of clitic-doubling constructions, especially in the configurations below.

(145)a. $[_{IP} [_{NP} x_i] [_{I'} CL_i + V]]$
b. $[_{IP} NP [_{I'} INFL [_{VP} CL_i + V [_{NP} x_i]]]]$

We have already seen that (145a) is excluded in Standard French when the [NP,IP] position is filled by phonologically realized material because the position is not Case-marked.

Now the sentences in (144) could tentatively be represented as in (146).

(146) $[_{IP} [_{NP} x_i] [_{I'} V + CL_i]]$

The only difference between (145a) and (146) is the position of the clitic, about which I will have nothing to say. We thus want to ask the following question: are the sentences in (144) instances of subject doubling? If the answer is positive, it then has to be followed by another question: why is (146) acceptable in Standard French given that (145a) is not? In other words, if we can have doubling in (146) why can we not have it elsewhere?

I argue here that the answer to the first question is negative and that this follows from the hypotheses already introduced in chapters 2 and 3. This position with respect to Complex Inversion is then shown to have desirable consequences for characterizing some of the properties of this construction.

Let us assume, as we have done so far, that St.Fr. subject clitics are base-generated in INFL, that they govern an instance of pro in [NP,IP], and that they absorb the Case feature [+C] of AGR. In other words, they participate in the formation of a c-chain. We then predict that no phonologically realized material can be present in [NP,IP] when this position is coindexed with a subject clitic in INFL. If there were such material, it would be made illicit by the Case Filter. This overt material (the lexical NP) must therefore occupy another position than that of the external argument in Complex Inversion. There is evidence in favor of this approach.

Note first that the presence of the lexical NP in Complex Inversion is optional.[32]

(147)a. Viendra-t-il? =(144a)
b. A-t-elle fait cela? =(144b)

The sentences in (147) are then the same as the ones in (148) with respect to the relationship between the subject clitic and the external argument position illustrated in (149). They differ in the surface position of the clitic.

(148)a. Il viendra.
b. Elle a fait cela.

(149)

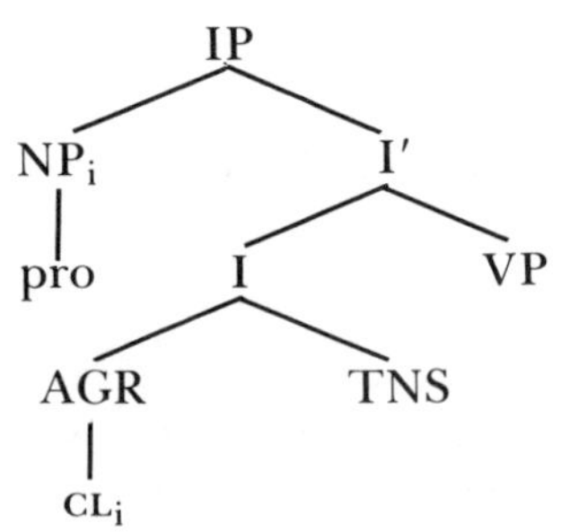

In Standard French, no lexical NP can replace pro in (149). Let us assume that this is also true of Complex Inversion and accept the fact that the lexical NP in (144a,b) is probably in a position external to IP as in (150).[33]

(150) $[_{NP} \ldots] [_{IP} [_{NP}\ pro_i] [_{I'} [_{INFL}\ CL_i \ldots] [_{VP} \ldots]]]$

The NP located outside of IP does not require Case-marking since it is not in argument position.[34] But what is its exact position? The only clear indication we can get is that it is not in a TOP node if this node is outside of CP as is often assumed, because the NP appears to the right of the *wh*-word in *wh*-questions as in (151).

(151)a. Quand Pierre viendra-t-il?
'When will Pierre come?'
b. Pourquoi Marie a-t-elle fait cela?
'Why did Marie do that?'

Kayne (1972), citing Gross (1968), mentions that a dislocated NP does not usually appear between the *wh*-word and the verb. This observation is based on the claimed contrast in (152); cf. also 3.1.3.1.

(152)a. Pourquoi est-ce mauvais, l' argent?
why is it bad the money
'Why is money bad?'
b. ?? Pourquoi, l'argent, est-ce mauvais?

I do not believe that this argument is sufficient to reject the analysis proposed here since, in my judgement, the sentences in (153) are all acceptable. I use the comma in the traditional way, which is not relevant here.

(153)a. Je me demande quand, Pierre, il va venir.
'I wonder when, Pierre, he will come.'
b. Je me demande quand il va venir, Pierre.
c. Je me demande, Pierre, quand il va venir.
d. Pierre, je me demande quand il va venir.
e. Pierre, quand va-t-il venir?
f. Quand va-t-il venir, Pierre?
g. Quand Pierre va-t-il venir?

There seems to be no reason why the NPs in (151) cannot be external to IP. In fact, according to our analysis, the NPs simply cannot be in external argument position; they would remain unmarked for Case in such a position. Let us therefore assume that the NPs in (144) are adjoined to IP. This makes them appear to the right of COMP in *wh*-questions as in (151) above.

This approach has three desirable consequences that I now present.

First, a lexical NP in Complex Inversion cannot undergo *wh*-movement.

(154)a. Jean viendra-t-il?
b. *Qui$_i$ viendra-t-il$_i$?
'Who will come?'
c. Ces personnes seront -elles présentes?
these people be + FUT-3P-CL present
'Will these people be present?'
d. *Quelles personnes$_i$ seront-elles$_i$ présentes?
'Which people will be present?'

This is a trivial matter if we assume that the NPs are not in subject position: they cannot undergo subject extraction. Alternatively, it can be assumed that in order to undergo subject extraction, the NPs must be in subject position. The sentences are then ungrammatical because the variable left by *wh*-movement of the subject is not Case-marked. The two options are illustrated in (155) and (156) below.

(155) [$_{CP}$ [$_{IP}$ NP [$_{IP}$ [$_{NP}$ pro$_i$] [$_{I'}$ . . .]]]]

(156) [$_{CP}$ *wh*$_i$ [$_{IP}$ [$_{NP}$ *t*$_i$] [$_{I'}$ . . .]]]
[-Case]

Note that (154b,d) are excluded, in fact, for the same reason the sentences in (157a,b) are, as opposed to (157c,d).

(157)a. *Qui$_i$ il$_i$ vient?
who CL come-3S
'Who comes?'
b. *Je connais la fille qu'$_i$ elle$_i$ chante.
I know-1S the girl that she sing-3S
'I know the girl who sings.'
c. Qui vient?
d. Je connais la fille qui chante.

Secondly, a subject NP in argument position in St.Fr. can undergo the well-known rule of Stylistic Inversion; cf. Kayne (1972;1975), Kayne and Pollock (1978), and Safir (1985) for discussions of this construction.

(158)a. Quand viendra Pierre?
'When will Pierre come?'
b. Quand a chanté Marie?
'When did Marie sing?'

If the NPs in (151) and more generally in Complex Inversion constructions are not in subject position, then it is predicted that they cannot undergo the rule of Stylistic Inversion. That this is true is illustrated in (159) where both Stylistic Inversion and Complex Inversion are found.

(159)a. *Quand viendra-t-il Pierre?
b. *Quand a-t-elle chanté Marie?

Thirdly, it can be noted that any type of NP can be found in Complex Inversion. The construction is not limited to non-quantified NPs as the Colloquial French subject doubling is as shown in (160); cf. 3.1.3.1.

(160)a. Chaque spectateur doit-il payer?
each spectator must-CL pay
'Must each spectator pay?'
b. Tout est-il en ordre?
all is -CL in order
'Is everything in order?'
c. Une table peut-elle avoir deux pattes?
a table can -CL have two legs
'Can a table have two legs?'

One could argue that the sentences above are indeed instances of subject doubling and that Standard French corresponds to a language in which only LF extractions out of doubling are allowed, as is the case in PN Fr. But this would bring us back to our original problem of determining why subject doubling is possible only in interrogatives in St.Fr.

Finally, it would be interesting to explore what happens in a language allowing subject doubling and exhibiting Complex Inversion. There would be no reason to exclude the presence of the NP in argument position in such a language. Colloquial French is not a good example of this type of language since what appears at first sight to be Complex Inversion is different enough to cast doubt on the claim that it is equivalent to St.Fr. Complex Inversion.[35]

Nevertheless, Trentino exhibits what is, according to Safir (1985, 221), a property very similar to St.Fr. Complex Inversion. Compare the following (a) sentences in Trentino to the (b) sentences in St.Fr.

(161)a. Magne*lo*?
b. Mange-t-*il*?
'Does he eat?'

(162)a. A*lo* magna?
b. A-t-*il* mangé?
'Has he eaten?'

(163)a. La magne*lo*?
b. La mange-t-*il*?
'Does he eat it?'

(164)a. *El$_i$ magne*lo*$_i$?
b. *Il$_i$ mange-t–*il*$_i$?

(165)a. Quando magne*lo*?
b. Quand mange-t-*il*?
'When does he eat?'

(166)a. El Mario magne*lo*?
b. Mario mange-t-*il*?
'Does Mario eat?'

It is hard to verify whether the lexical NP in (166a) is in argument position or not. Recall that one of the crucial reasons why we assume that it is not in (166b) is because this NP cannot be moved either in *wh*-questions or in Stylistic Inversion constructions. This is so because even if the NP was in [NP,IP] position in Trentino, its extraction in syntax would be prohibited as we have seen in 3.3.1.4. Extraction out of doubling constructions is blocked in this dialect. The sentences in (167) are ungrammatical.

(167)a. *Chi$_i$ *el*$_i$ ha magna?
b. *Chi$_i$ ha*lo*$_i$ magna?

Furthermore, Stylistic Inversion is blocked with the presence in the sentence of a coreferential inverted subject clitic. As in St.Fr., the sentence is acceptable only under a right-dislocation interpretation with a pause between the verb and the postverbal NP.

(168)a. Quando magne*lo* *(##) el Mario?
b. Quand mange-t-*il* *(##) Mario?

The only valid conclusion at this stage is that it is unclear whether the lexical NP in (166a) is in [NP,IP] or in a position external to IP, with pro in [NP,IP]. The two analyses are plausible ones. I will not pursue this matter further.

In conclusion, our theory on the relationship between a clitic and its associated argument position as it applies in Standard French is supported by the Complex Inversion facts if we assume that a lexical NP involved in such a construction is in a position external to IP, probably adjoined to it.

Note however that the proposed structure in (150) above requires that pro in [NP,IP] – coindexed with and governed by the subject clitic in INFL – also be associated with the NP external to IP, probably by coindexation. This violates (136iii) and should make it impossible for the clitic to form a c-chain with pro in [NP,IP]. As a consequence condition (137) and ultimately the Recoverability Principle would be violated.

More generally, we do not want (136iii) to rule out empty resumptive pronouns in general. In other words, it must be assumed that (136iii) only refers to an element $\bar{A}$-bound as a result of movement. In that case, (150) would not cause problems for (136) and (137).

3.5. A NOTE ON REFLEXIVE *se*

In this section I wish to repeat the exercise done in the previous section with regard to Complex Inversion, concentrating this time on reflexive clitic *se* in French. Note however that this section is, at best, incomplete in many respects. An exhaustive presentation of the various properties of reflexive and reciprocal French *se* is available in Kayne (1975).[36]

The problem posed by sentences like the ones in (169) for our analysis of clitics has to do with the status of the empty category related to the reflexive clitic.

(169)a. Jean se voit.
CL(REFL) see-3S
'Jean sees himself.'
b. Jean se lave.
'Jean washes himself.'

To see this, let us explore one possible analysis. Suppose that in order to capture the fact that the reflexive is anaphoric to the subject, we say that the empty category in internal argument position is coindexed with the NP in external argument position as it presumably is in the English translations in (169). We would get something like:

(170)a. $[_{IP}[_{NP}\ \text{Jean}_i]\ [_{I'}\ \text{INFL}\ [_{VP}[_V\ \text{se}_i + \text{voit}]\ [_{NP}\ e_i]]]]$
b. $[_{IP}[_{NP}\ \text{Pierre}_i]\ [_{I'}\ \text{INFL}\ [_{VP}[_V\ \text{se}_i + \text{lave}]\ [_{NP}\ e_i]]]]$

The empty category in (179a,b) is clearly [+anaphor] according to the Binding Conditions. In fact, it cannot be anything else since it is bound (by the subject) in its governing category (IP).

With respect to the clitic, there are two possibilities. The reflexive clitic is either coindexed with the empty object NP or it is not. That is where the problem lies. Following the first option, the empty category has to be pro because of the conditions established earlier on c-chains (cf. 3.3.4) but then, still according to the definition of c-chain, the empty category must be locally (A or $\bar{\text{A}}$)-free. There is thus a contradiction: the clitic must license a [+pronominal, −anaphor] empty element in internal argument position but in order to be assigned a reflexive reading the empty category in (169) must be [+anaphor].

The second option seems at first sight even more problematic. Suppose that the reflexive clitic is not coindexed with the internal argument (the object). This allows for the desired reflexive interpretation insofar as the subject and the object are coindexed and, thus, coreferent. However, the adoption of such an approach could translate as the loss of the generalizations made up to now with regard to (object) clitics. We would lose in particular the condition (137) above which was argued for independently. One would have to propose a seemingly ad hoc solution stipulating that (137) does not apply to reflexive clitics. This seems rather undesirable.

On the other hand, there is evidence that supports such an approach. Let us then assume that reflexive *se* is not coindexed with an argument position and see how this can be implemented.

Let us return to the basic intuition that led us to formulate condition (137) on the appearance of clitics. Very generally, a clitic fulfills the function of recovering some information referring to null syntactic arguments. Such information includes person, number, Case, and maybe gender features. A clitic carries or instantiates this information as features which are phonologically realized in a certain way depending on the particular combination of features in question.

In Standard French, this realization of the features comprises the ones in (171).

(171)a. [3, sg, masc, accusative] = le (je *le* vois)
b. [3, sg, masc, nominative] = il (*il* voit)
c. [3, sg, masc/fem, dative] = lui (je *lui* parle)
d. [3, pl, masc/fem, accusative] = les (je *les* vois)
e. [2, sg, masc/fem, accusative] = te (je *te* vois)

In a way similar to rich agreement in Italian, clitics in Standard French vary in their form for different feature combinations. Reflexive *se* does not pattern with the other clitics in many relevant respects.

First, the same phonological realization serves to license both accusative and dative reflexive objects. Compare (172) with (173).

(172)a. Jean lui parle.
CL(DAT) speak-3S
'Jean speaks to him.'
b. Jean le voit.
CL(ACC) see-3S
'Jean sees him.'

(173)a. Jean se parle.
DAT
'Jean speaks to himself.'
b. Jean se voit.
ACC
'Jean sees himself.'

Secondly, *se* does not vary in gender as illustrated by the contrast between (174) and (175).

(174)a. Jean le voit.
CL(MASC)
'Jean sees him.'
b. Jean la voit.
CL(FEM)
'Jean sees her.'

(175)a. Jean se voit.
MASC
'Jean sees himself.'

b. Marie se voit.
FEM
'Marie sees herself.'

Thirdly, *se* is invariable as far as number is concerned; other clitics are not.

(176)a. Jean le voit.
CL(SG)
b. Jean les voit.
CL(PL)
'Jean sees them.'

(177)a. Jean se voit.
SG
b. Les gens se voient.
the people PL see-3P
'The people see themselves.'

Finally, reflexive *se* is exclusively a third person clitic. The usual object clitics are used for first and second person.

(178)a. Je me vois.
1S
'I see myself.'
b. *Je se vois.
c. Tu te vois.
2S
'You see yourself.'
d. *Tu se vois.
e. Nous nous voyons.
1P
'We see ourselves.'
f. *Nous se voyons.
g. Vous vous voyez.
2P
'You see yourselves.'
h. *Vous se voyez.

We see, therefore, that as far as the feature composition of *se* is concerned, no phonological variation is involved, contrary to the case of regular clitics. This is the first piece of evidence in favor of a principled distinction between *se* and the other clitics.

The existence of the so-called "inherent" *se* can also be taken as evidence for a different status for this clitic. Consider the sentences in (179).

(179)a. Il s'en va.
'He is leaving.'
b. Paul s'évanouit toujours à la vue du sang.
'Paul always faints at the sight of blood.'
c. Jean s'efforce de bien faire.
'Jean strives to do well.'
d. Le château s'est écroulé.
'The castel tumbled away.'
e. Les enfants se chamaillent.
'The children squabble.'

In these sentences there is no argument position for the reflexive to relate to, as noted in Kayne (1975, 386).[37] It can be assumed simply that the clitic *se* is part of the lexical entry of these intransitive verbs. In fact, most of them do not exist independently of *se*.

(180)a. *Il en va (lui(-même)).
him(self)
b. *Paul évanouit (lui(-même)).
c. *Jean efforce (lui(-même)) de bien faire.
d. *Le château a écroulé (lui(-même)).
e. *Les enfants chamaillent (eux(-mêmes).
them(selves)

Finally, Borer (1984, 123) notes the very interesting fact that there is complete absence of clitic doubling with reflexive clitics in the languages which otherwise allow such doubling of objects. She refers in particular to Romanian and River Plate Spanish. This cluster of properties specific to reflexive *se* and similar reflexives in other Romance languages must be accounted for.

The fact that *se* does not vary in its surface form for different feature combinations suggests that it does not fulfill the same function as other clitics, namely that of recovering missing syntactic information. Intuitively, there is an obvious reason for this state of affairs. If it is true that the NP object position has the properties of an anaphor, then it must be bound in its governing category and the only available A-binder is the subject. In other words, everything works as if the subject itself was sufficient for the recoverability of the missing object. Therefore, condition (136), designed to account for the recovering function of clitics, is not needed for reflexive *se* simply because there is no missing information.

This preliminary conclusion finds support in the existence of inherent *se*. The sentences of (179) – and there is a wealth of other similar verbs in Standard French[38] – are the perfect illustration of *se* not even being able to enter into the formation of a c-chain since the argument position necessary for this process to happen is not available.

Following Borer (1984), we take the absence of doubling with reflexive *se* as an argument for the anaphoric status of the empty object NPs of reflexive verbs. Borer (1984, 122) writes: "Such an NP will be both an R-expression, which has to be free, and an anaphoric expression, by virtue of the particular interpretation [of the sentences]."

It is now possible to make a concrete proposal. I assume the analysis of Borer (1984), who suggests the following representation for clitic *se*.

(181) $[_{VP} [_{V} \begin{bmatrix} \text{se} \\ \text{i} \\ \Theta_2 \end{bmatrix} + V] [_{NP}\ e_i]]$

This representation simply illustrates the fact that *se* is not coindexed with the object NP but that it is associated with the Θ-role (in the Θ-grid of V) assigned to that NP. The representation for regular clitics is given in (182), where *cl* represents a combination of the features [person, number, gender, Case].

(182) $[_{VP} [_{V} \begin{bmatrix} cl_i \\ \text{i} \\ \Theta_2 \end{bmatrix} + V] [_{NP}\ e_i]]$

The representation in (181) is specified as anaphoric. This forces the [+anaphor] empty category in [NP,VP] to be bound by the subject. The latter acts as the element necessary for the recoverability of the features of the missing object. Note, finally, that the empty NP cannot be PRO, the pronominal anaphor since it is governed by the verb.

In conclusion, reflexive *se* does not act as a clitic with respect to condition (137) for the reasons stated above. The only characteristic *se* shares with other clitics is that it absorbs Case.

3.6. SUMMARY

The main assumption that guided our analysis of the doubling phenomenon was that there exists in the grammar a concept "clitic" which encompasses both subject clitics and object clitics. Under this view, both types of clitics are seen as two distinct syntactic realizations of a same element. In chapter 2, it was argued that this assumption could account in a simple manner for the symmetry existing between subject clitics and object clitics in the way they license pro in the argument position to which they are related.

A natural consequence of this particular view was that instances of doubling should be analyzed in a unified way whether they involved subject clitics or object clitics. This was done in 3.1 and 3.2 where I suggested a formal approach to Case-absorption and developed a particular way to describe what is involved in the assignment of Case to clitic-doubled overt NPs.

In 3.3, I have examined the extraction possibilities out of doubling constructions in seven languages/dialects and concluded that these possibilities should be handled by a parameter referring to the level at which a clitic and the related argument enter in the formation of a c-chain. The definition of this particular type of chain gathers all the conditions on the appearance of clitics previously examined. One of the effects of this definition is to block any extraction of an NP that is coindexed with and governed by a clitic. Extractions are permitted or barred in certain languages/dialects. When they are allowed, they can happen either in syntax or at LF or both but never only in syntax. This was taken as evidence that a parameter involving the level at which c-chains can be broken is involved. Confirmation for this type of analysis comes from Pied Noir French which exhibits both subject doubling and object doubling. The restrictions on extraction from these two constructions are exactly parallel and support the unified analysis proposed here.

In the last two sections of this chapter, two potentially problematic constructions were briefly examined. In 3.4, we were concerned with the apparent subject-doubling phenomenon found in French Complex Inversion and concluded that there is, in fact, no doubling involved there. This was made possible by the assumption that a lexical NP subject in Complex Inversion can be viewed as being generated outside of IP, with pro in [NP,IP] licensed by the inverted subject clitic. In 3.5, I discussed some problems related to reflexive *se* in French and argued that it is not part of a c-chain. This is, in turn, interpreted in core grammar as characterizing an anaphoric relation between the external argument and the internal argument.

The following chapter is devoted to sòme residual but important issues raised by the discussions in chapters 2 and 3.

CHAPTER FOUR

Clitics and Agreement Markers

4.0. INTRODUCTION

It is a well-known fact that weak forms of subject and object pronouns in French, Trentino, Fiorentino, Pirahã, and other languages are cliticized onto the inflected verb of their sentence. It was argued, especially in chapter 2, that subject clitics are generated in AGR of INFL where they serve the function of licensing an empty pronominal pro in external argument position. In this sense, they serve the purpose of recovering the identity of the missing information represented in syntax by the empty argument. We have also seen that this observation holds for the internal argument position associated with an object clitic, which is affixed to the inflected verb. For this reason, we found it useful to subscribe to the hypothesis that subject clitics and object clitics are two different realizations of a same syntactic element. The structural differences between internal and external arguments proved to be a good basis for characterizing the differences between their associated clitics.

There is a further distinction to be made between subject clitics and object cliitcs; it has to do with the way in which they achieve their status as verbal affixes. We have assumed throughout the present work that object clitics are base-generated on the verbal head. Obviously this is not true of subject clitics since their base position is somewhere in AGR of INFL.

In the first section of the present chapter, the different options that arise in order to account for the clitic status of the subject pronouns are explored. I concentrate on French but the analysis adopted here could be extended to other languages with subject clitics and having syntactic structures similar to the ones found in French.

The second section is concerned with the status of clitics with respect to agreement or inflectional markers. I wish to offer brief comments on the differences between the Romance type of languages and the so-called concatenative or agglutinative languages with regard to the null argument property and the doubling phenomenon.

4.1. SUBJECT CLITICIZATION

Let us return to Chomsky's (1981) account of the null subject property. His analysis relies on the fact that the element INFL must be attached to the verb at some point in the derivation. This is true simply because the inflectional features (person, number, gender, and tense) are affixed to the verb on the surface. Chomsky chose the option of moving INFL to VP in order to express this fact. This is what he characterized as rule R in his account of the null subject property; cf. 1.2.1.

One of the reasons underlying this choice is the need for the external argument position in null subject languages to be ungoverned in order to allow it to be occupied by PRO. Hence, if rule R (Affix Hopping) applies prior to S-structure, PRO is ungoverned when the Binding Conditions apply to the structure and it follows that an empty subject can surface in the languages where rule R applies at this particular level. In other languages, Phonological Form is the relevant level for the application of rule R and null subjects are consequently barred. Another reason for this choice is the relationship that was believed to hold between null subjects and free inversion. INFL in VP is in a configuration in which [+TNS] can assign nominative Case to an NP adjoined to VP.

Now according to the various analyses presented in the previous chapters, in particular section 2.4 and chapter 3, the following remarks must be made. First, the empty category associated with null subjects is no longer taken to be PRO in tensed clauses. Following Chomsky (1982), we have seen that pro is best suited to occupy this position. There is thus no need to eliminate government of the subject position by the INFL node. In fact, this type of government seems to be required for the licensing of pro by AGR, whether AGR contains a clitic or not; cf. 2.2. and 3.3.4. It follows that the rule affixing INFL to the verb to be inflected does not need to apply prior to S-structure. Again, it seems, in fact, as if it must not apply before S-structure given the well-formedness condition in (137) in chapter 3 on the appearance of clitics in the syntactic structure and its level of application, namely, LF in the unmarked case and

S-structure in the marked case. Furthermore, there is no difference between, say, French and Italian with respect to this rule. The difference between these two languages lies rather in the way each language licenses pro in subject position. French has recourse to clitic pronouns whereas rich agreement is sufficient for the licensing of pro in Standard Italian. Thus the first argument for movement of INFL to VP does not hold anymore.

Second, it was suggested in 2.4 that the free inversion property is made possible by the presence of the TNS element in VP prior to S-structure, where it can assign Case in the government domain of V. This was made necessary as the null argument property as it is conceived of here cannot be directly related to free inversion; cf. 2.3. The second argument for INFL-movement to VP, which is based on the relationship between null subjects and free inversion does not go through.

I propose to approach the problem of the affixal nature of subject clitics in French by assuming that the process which accomplishes this is the same one that affixes inflectional features to the verb. Two possible analyses arise: INFL/AGR moves to V (Chomsky 1981) or the verb to be inflected raises to INFL (Emonds (1978) and McA'Nulty (1983)). Before exploring the two options, we will briefly examine the internal structure of INFL.

4.1.1. The Internal Structure of INFL

Let us adopt the hypothesis that the internal structure of INFL in French must reflect the relative surface order of its constituents. The constituents of INFL are, among others, AGR, TNS, subject clitics, and the negative marker *ne*. The basic surface ordering, including the finite verb and its object clitics, is given in (1a).

Object clitics are base-generated on the finite verb. The verb, being in VP at D-structure, must be excluded from the structure of INFL at this particular level of representation. A plausible structure for INFL is the one given in (1b), where the subject clitic and *ne* are optionally generated.

(1)a. Subject Clitic + *ne* + [Object Clitics + V] + {TNS, AGR}
b.

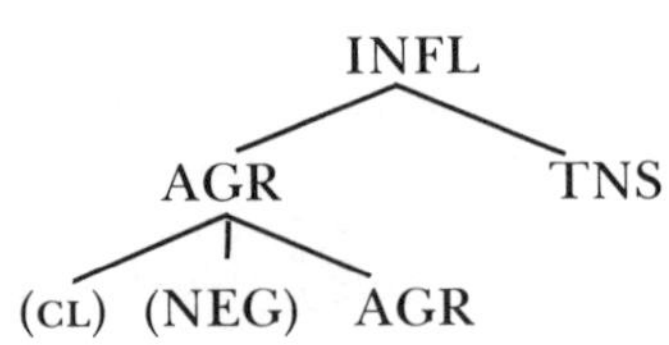

Note that this structure is specific to Standard French inasmuch as there is a great deal of cross-linguistic variation with respect to the negation marker, for instance; see Rizzi (1986a) for a discussion of the various positions that can be occupied by negative particles in various Italian dialects. In Trentino, the subject clitic always follows the negative particle (*no*). In Romagnolo, the negative particle follows the subject clitic as in Standard French. In Colloquial French there simply is no such negative marker in INFL; there is only *pas* adjoined to VP. Note also that (1b) indicates that clitics and AGR are different elements. The clitic, in a sense, is in a Spec AGR position and this allows the feature-agreement to show up independently on the verb.

4.1.2. Verb Raising or INFL Lowering?

McA'Nulty (1983) presents some arguments in favor of a rule of Finite Verb Raising (henceforth, FVR) as proposed in Emonds (1978). The effect of such a rule is illustrated in (2). "OCL" stands for "object clitic(s)".

(2) FVR:

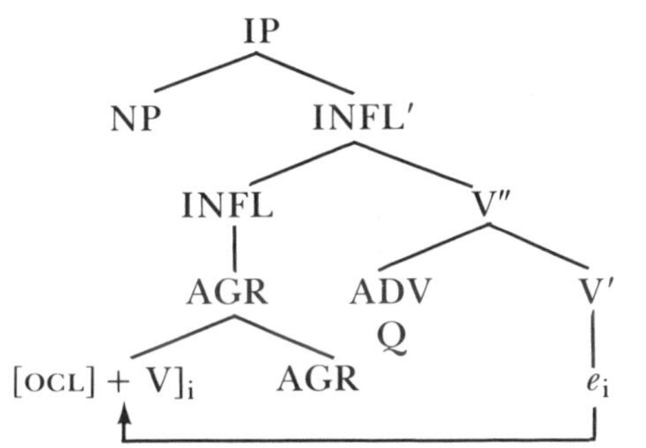

The rule moves the inflected verb together with its object clitics to a position adjoined to the left of AGR. The account just presented has the advantage of explaining the relative position of the clitics with respect to the agreement markers as in (3), where the relevant items are underlined.

(3) Pierre le lui donnera.
CL CL give-FUT-3S
'Pierre will give it to him/her.'

Consider next the sentences in (4)-(6) from McA'Nulty (1983, 139) and the different positions of the underlined elements.

(4)a. Jean préfererait ne pas les avoir mangés.
prefer-COND-3S NEG not CL to + have eaten
'Jean would prefer not to have eaten them.'
b. Jean ne les a pas mangés.
CL have-3S not
'Jean has not eaten them.'
c. *Jean ne pas les a mangés.

(5)a. Jean dit souvent les avoir mangés.
say-3S often CL to + have eaten
'Jean says to have often eaten them.'
b. Jean les a souvent mangés.
CL have-3S often
c. *Jean souvent les a mangés.

(6)a. Jean voudrait tous les avoir rencontrés.
want-COND-3S all CL to + have met
'Jean would want to have met all of them.'
b. Jean les a tous rencontrés.
CL have-3S all
'Jean has met all of them.'
c. *Jean tous les a rencontrés.

If the adverbs and quantifiers (*pas*, *souvent*, *tous*) in (4)-(6) are generated directly under V″ as in (2) then the contrasts are accounted for. The infinitive verbs in the (a) examples follow these elements because the verb stays in its base position, as the head of V″. However, a finite verb gets its inflectional features through FVR. On the surface then, such a verb precedes the adverbs and quantifiers. The contrast between the (a) and (b) sentences is explained, as well as the ungrammaticality of the (c) examples where a finite verb follows the adverbs and quantifiers.

The second plausible approach is Chomsky's (1981) rule R, called here INFL Lowering. The operation is illustrated in (7).

(7) INFL Lowering:

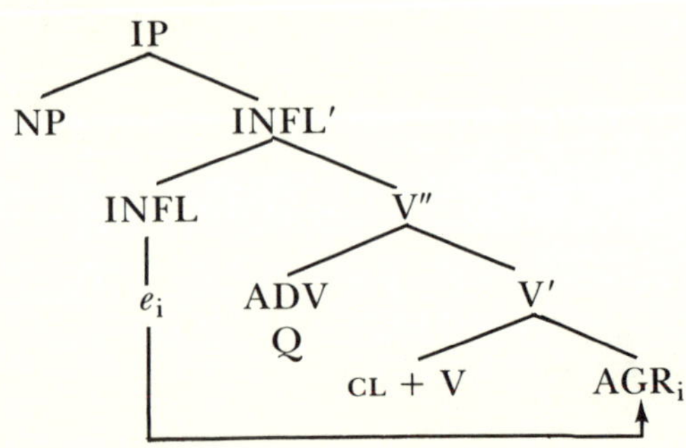

The respective position of the clitics and agreement markers illustrated in (3) is accounted for, as is the case for FVR. On the other hand, the contrasts in (4)-(6) are left unexplained unless the structural position of the quantifiers and adverbs is revised and modified. The prediction made by the rule in (7) is that a verb (tensed or not) always follows the adverbs and quantifiers. This is an undesirable prediction as shown in (4)-(6).

The two analyses presented above do not take subject clitics into account. The issue is as follows. In both the French and Italian dialects under study here, a subject clitic always appears as the first member to the left of the surface clitic string as shown in (8).

(8) *Il* le lui donnera.
CL CL CL give-FUT-3S
'He will give it to him/her.'

This causes problems in both analyses assuming that subject clitics are generated in AGR. Both FVR and INFL Lowering place the subject clitic to the right of the verb. It is easier to modify the FVR rule in order to overcome this problem. A possible modification to (2) is suggested in (9), which is consistent with the internal structure of INFL adopted above.

If we assume, following Chomsky (1986b, 4), that the two types of movement possible under Move α are substitution and adjunction, then the [OCL + V] complex can only be adjoined to AGR as in (9).

(9)

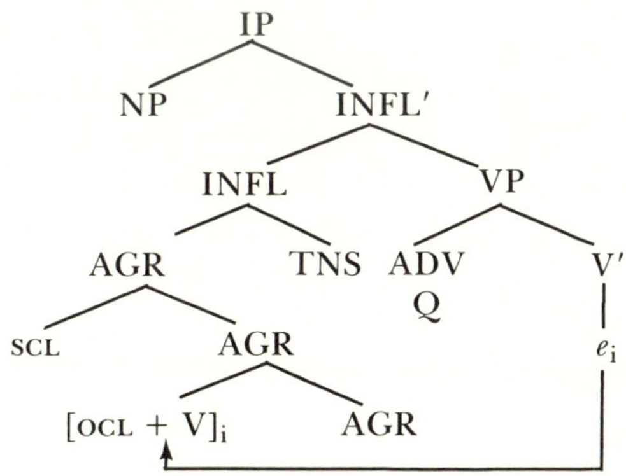

Both the relative ordering of the clitics in the string and the position of adverbs and quantifiers are accounted for. Furthermore, the inflectional features included in AGR and the tense features will end up as suffixes on the verb.[1]

FVR necessarily applies late in the derivation. The condition in (137) in chapter 3 forces it to apply in Phonological Form. There seems to be no evidence that this rule should apply in the syntactic component of the grammar. In that sense, the rule assumed here seems to differ from Chomsky's (1986b, section 11) V-raising which is a local head-to-head movement of a syntactic nature.

It is interesting to note that FVR as it is conceived of here has further desirable consequences in the grammar of French. Lobeck (1985;1986) proposes a similar rule and argues that it can explain the absence of VP anaphora or ellipsis in French. In this language, a sentence like (10) is ungrammatical.

(10) *Marie a vu le film et Paul a ϕ aussi.
'Marie has seen the film and Paul has ϕ too.'

Lobeck argues that the ECP mediates the occurrences of empty VPs in the sense that the empty VP requires a proper governor. There is such a governor in English where, it is argued, auxiliary verbs are in INFL during the syntax where they can serve as proper governors for empty VPs. In French, on the other hand, auxiliary verbs and main verbs both behave in the same way, as argued in Kayne (1975), and undergo FVR in PF. There is thus no proper governor for empty VPs at S-structure or at LF, whichever level the ECP applies at.[2]

The modified rule of Finite Verb Raising in (9) therefore allows us to account for a variety of phenomena, at least in French.

4.1.3. Coordination

In 2.2.4 we saw that Rizzi (1986a) appeals to certain differences between Trentino and Standard French in the behavior of subject clitics with respect to coordination to support the existence of two different types of subject clitics.[3] Recall the sentences in (11) and (12).

(11) Standard French
Elle chante et danse.
'She sings and dances.'

(12) Trentino
a. La canta e la balla.
'She sings and she dances.'
b. *La canta e balla.

According to Rizzi, a zero pro-form is barred in AGR under coordination; see 2.2.4. Therefore, by assuming base generation of the subject clitics in AGR in Trentino, he can provide an explanation for the ungrammaticality of (12b) and the obligatory repetition of a subject clitic in the second conjunct as in (12a). On the other hand, if French subject clitics are in [NP,IP] and later cliticized to the verb, then (11) is predicted to be grammatical since there is no zero pro-form in the AGR node of the second conjunct.

In this section, I offer another interpretation of the coordination facts following which Rizzi's conclusion is reevaluated. This analysis is based on three premises: 1) subject clitics are base generated in AGR of INFL in the languages exhibiting them; 2) FVR applies late in the derivation and subsumes cliticization; and 3) coordinate structures are representations available during the syntax and are subject to syntactic rules. In the remainder of the section, I simply present the analysis proposed in Lobeck and Roberge (1985).

In Colloquial French, a lexical NP may be followed by two coordinated VPs as in (13).

(13) Marie danse pis chante.
'Marie dances and sings.'

When the subject is represented by a clitic the grammaticality of the sentence is questionable, as in (14), unless the clitic is repeated in both conjuncts, cf. (15).[4]

(14) ??Elle danse pis chante.

(15) Elle danse pis elle chante.

Let us tentatively assume the structure in (16) for (15), where subject clitics are generated in their usual position and INFL' projections are coordinated.

(16)

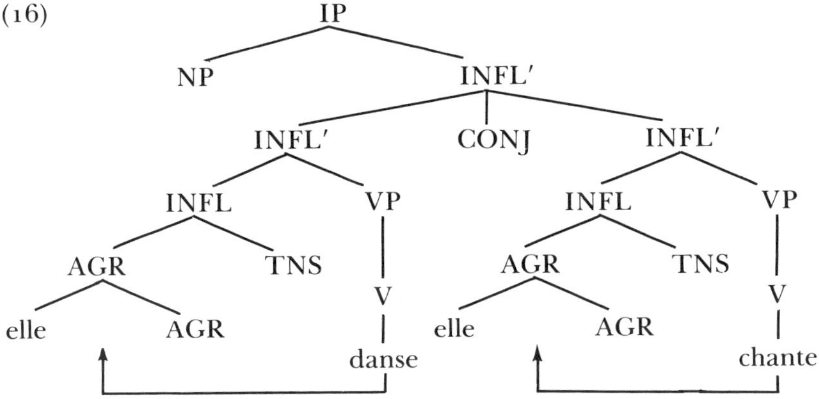

Consequently the structure of (13) is as in (17).

(17)

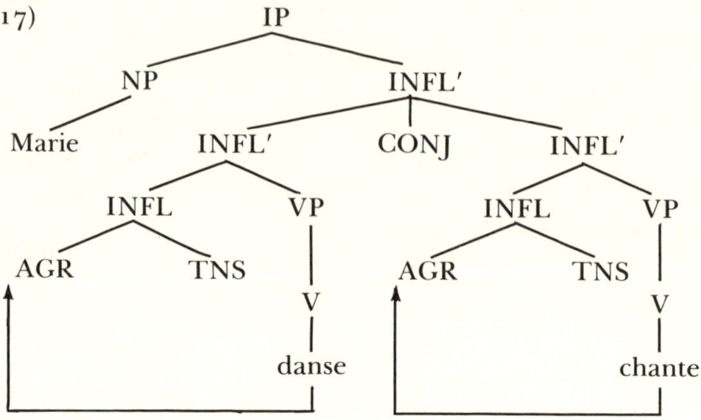

An alternative analysis could be suggested. Suppose that (13) is actually a coordination of two VP projections as illustrated in (18) instead of a coordination of INFL′ projections as in (17) above.

(18)

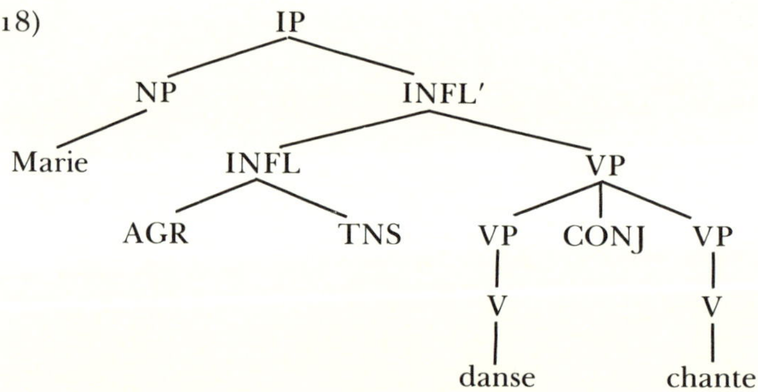

This hypothesis causes a problem for sentences like the one in (19), where VPs dominating forms of the auxiliary verbs *avoir* and *être* are coordinated. Following Kayne (1975), it is assumed that auxiliary verbs in French are dominated by VP.

(19) Marie a mangé pis est allée voir un film.
has eaten and is gone to + see a film
'Marie ate and went to see a movie.'

(20)

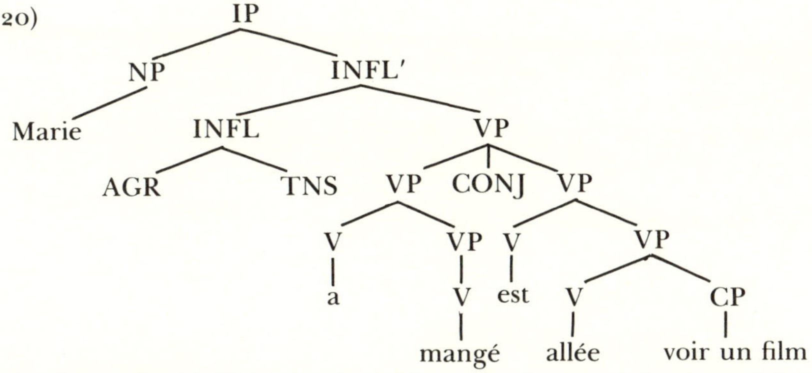

In Phonological Form, FVR will apply in both (18) and (20). The leftmost verb in each VP is raised to INFL as in (21a) and (22a), generating the ungrammatical sentences in (21b) and (22b).

(21)a.

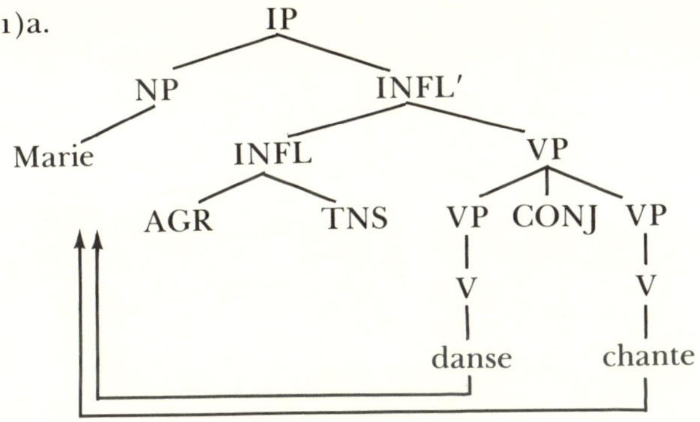

b. *Marie chante danse pis.

(22)a.

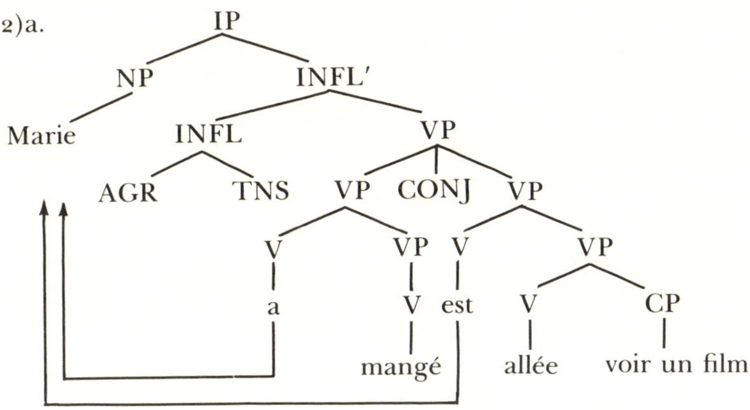

b. *Marie a est mangé pis allée voir un film.

Besides generating ungrammatical strings, the operation in (21) and (22) violates Ross's (1967, 89) Coordinate Structure Constraint, given in (23), since two verbs have raised out of conjoined VPs into INFL.

(23) Coordinate Structure Constraint
In a coordinate structure, no conjunct may be moved, nor may any element contained in a conjunct be moved out of that conjunct.

On the other hand, FVR applying in a structure with coordinated INFL′ does not violate the Coordinate Structure Constraint and produces the right outputs, i.e. (13) and (19) with the structures in (24) and (25).

(24)
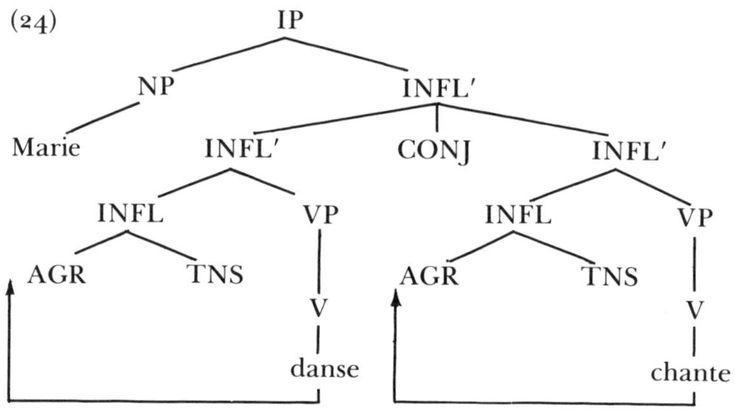

(25)
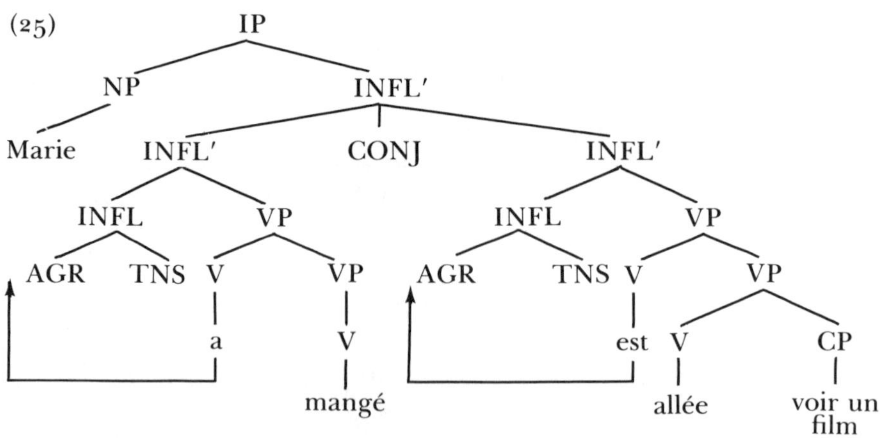

So what appears to be coordinated VPs on the surface is in fact coordinated INFL′ projections in French. It is only in coordinated INFL′ structures that FVR may apply without violating the Coordinate Structure Constraint.

Sentences like the one in (26) are also possible in Colloquial French where one auxiliary verb is followed by coordinated VPs, in this case the two past participles *dansé* and *chanté*.

(26)	Marie	a	dansé	pis	chanté.
		has	danced	and	sung

'Marie danced and sang.'

A D-structure for (26) where two VPs are coordinated is blocked for the reasons just outlined: it would require a structure where the finite auxiliary verb *a* is base-generated in INFL instead of VP.

(27) *

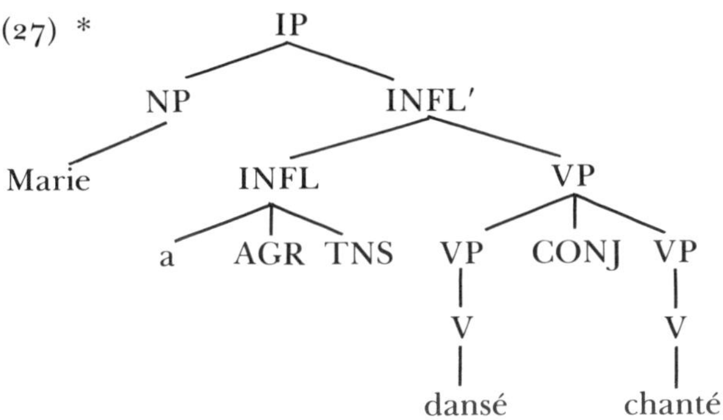

The structure in (28), where INFL′ projections are coordinated is also an impossible source for (26); there is no auxiliary preceding the past participle in the second conjunct.

(28) *

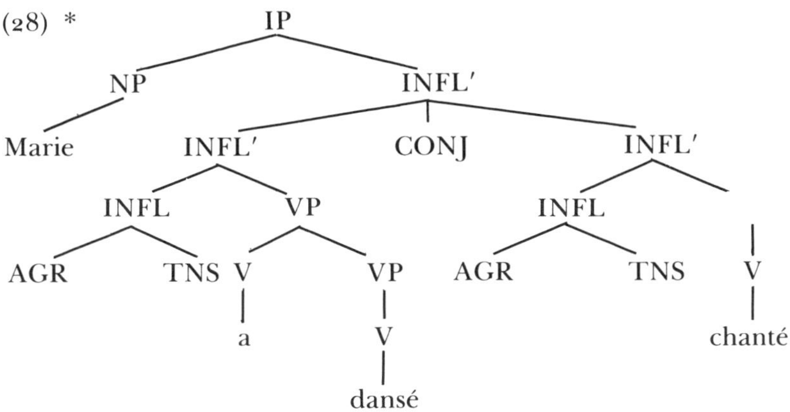

We are able to generate sentences like (26) by adopting an exception clause to the Coordinate Structure Constraint; the Across-the-Board Constraint in (29), which is proposed by Ross (1967) and discussed extensively in Williams (1978).

(29) The Across-the-Board Constraint
The Coordinate Structure Constraint may not be violated unless the same element is moved out of all the conjuncts.

The Across-the-Board Constraint allows restricted violations of the Coordinate Structure Constraint. Elements may be moved out of a coordinate structure only when such elements are identical. As a result of Across-the-Board application of a movement rule, the two identical elements moved out of coordinate structures appear as a single element in the surface string. This is schematically illustrated in (30).

(30) Across-the-Board Rule Application

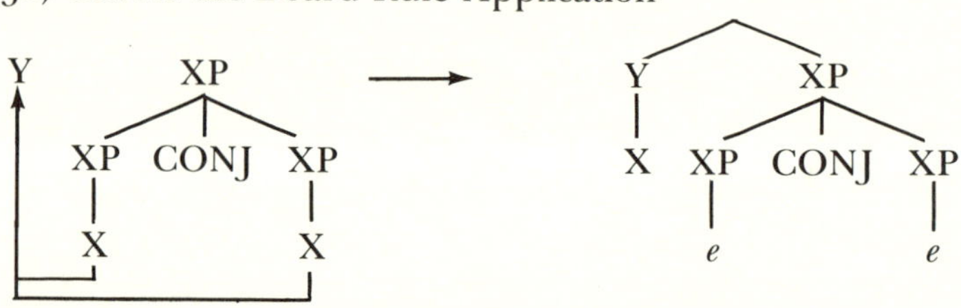

The sentence in (26) can be generated by allowing FVR to apply across-the-board to both conjuncts, raising identical auxiliary verbs out of each VP into INFL as in (31). This results in a surface string where only one auxiliary is phonologically realized.

(31)

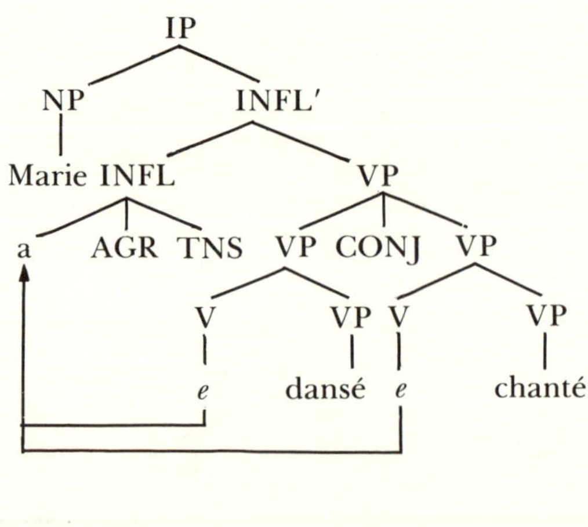

The interaction of FVR and constraints on rule application in coordinated structures predicts that in French, structures where VPs are coordinated are allowed only when FVR is able to apply across-the-board.

We are now able to generate sentences with subject clitics in Colloquial French in a straightforward way. In the grammatical sentence in (15) the conjuncts contain non-identical verbs. Because FVR is blocked from applying across-the-board, it must apply individually in each conjunct. We conclude then that INFL′, not VP, is coordinated. The subject clitic *elle* is generated in AGR in INFL in each conjunct and it cliticizes to the verb after FVR applies as illustrated in (32).

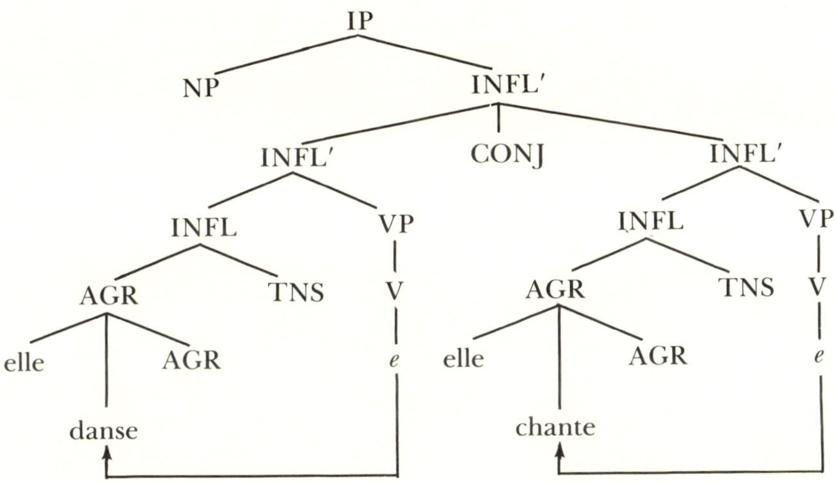

We are also able to generate the grammatical sentence in (33), where the subject clitic and the auxiliary verb *a* are followed by a coordinated VP.

(33) Elle a dansé pis chanté.
she have-3s danced and sung
'She danced and sang.'

The D-structure for (33) is given in (34), where two VPs containing the auxiliary verb *a* are coordinated. FVR applies across-the-board to raise the identical auxiliary verbs into the single INFL node dominating the subject clitic.

(34)

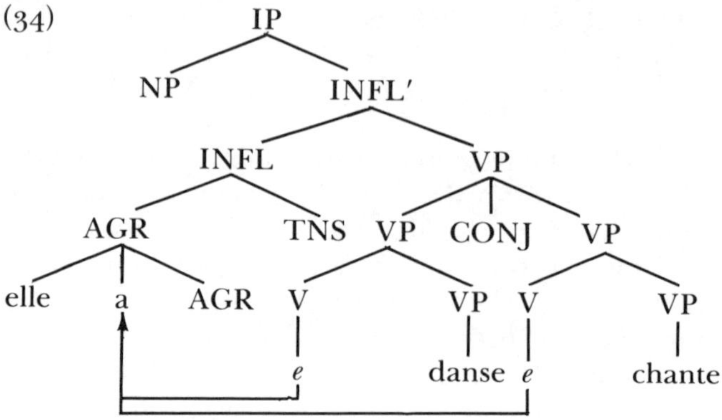

The questionable grammaticality of (14), where one subject clitic is followed by conjoined VPs, is explained under our hypothesis. The subject clitic *elle* may be generated in INFL, but the derivation will be ruled out because FVR violates the Coordinate Structure Constraint. This is illustrated in (35).

(35) ??

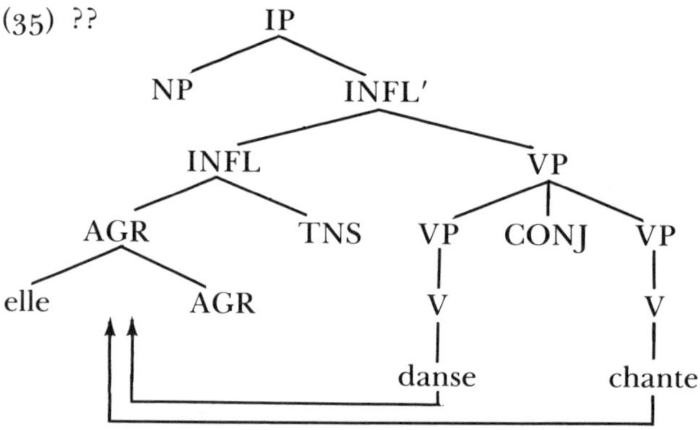

In conclusion, the properties of the French finite auxiliary and main verb interact with constraints on rule application in coordinate structures to allow us to predict whether INFL′ or VP is coordinated. We are then able to provide an account for the distribution of subject clitics in Colloquial French in coordinate structures which supports the analysis of subject cliticization in (9).

Now, with respect to the ungrammaticality of (12b) in Trentino, Safir (1985, 350) alludes to the possibility that there may be a difference between the properties of conjoined structures in this dialect and those in French. We follow this intuition and assume that since subject clitics are obligatorily generated in AGR in INFL in Trentino, it is only possible to coordinate INFL′ projections. The only possible structure in Trentino is therefore equivalent to the one in (32). The other possibility, namely the one in (35), is blocked independently as argued above.

We can thus conclude that the difference in the behavior of subject clitics in coordination between French and Trentino is not due to different structural positions for the elements in the two languages but follows instead from various constraints on coordinate structures and the fact that subject clitics are obligatory in Trentino.

4.2. AGREEMENT MARKERS VS. CLITICS

I now turn to a rather different, although related, topic. McCloskey and Hale (1984, 525) state that "Agreement and Cliticization [. . .] look simply like different morphological reflections of a single syntactic configuration, one in which pro occurs in an argument-position governed by a lexical head specified for person and number features." But are cliticization and agreement the same in every respect? This issue is unfortunately too vast to allow me to discuss it extensively and thus we will be satisfied with a statement of the problem and some suggestions as to how to approach it.

To illustrate the problem more clearly, consider the sentences below where SCL = subject clitics, CL = other clitics, SA = subject agreement marker, and OA = object agreement marker.

(36) French
Il lui en a offert.
SCL + CL + CL + have-3s + offered
'He has offered some to him/her.'

(37) Italian
Me lo dice.
CL + CL + tell-3s
'He tells it to me.'

(38) Swahili (Barrett-Keach 1980)
a li mw ona
SA + PAST + OA + see
'He saw him.'

(39) Chilcotin[5]
be ray s tæl
OA + ASP + SA + kick
'He kicked me.'

(40) Pirahã (Everett 1985)

hi	hi	xib -áo	-b	-á
SCL +	CL +	hit + TELIC +	ASP +	MOOD

'He hit himself/him.'

What these few examples show is that sentences in these languages can be composed of a single verb but still include object(s) and/or a subject. The objects and subject are expressed by the presence of affixes on the verb. It is therefore legitimate to describe the French and Italian sentences in (36) and (37) as having a representation similar to the one in (41) as far as the verb is concerned.

(41) $[_V$ CL – CL – CL – V]

In (41), CL stands for an empty slot in V that can be filled by overt material representing the subject in French and a variety of objects in both French and Italian.

Now, still in French and Italian, there is something more that is affixed to the verb. In this case, there is often a suffix also representing, in some sense, the grammatical subject of the sentence. Representation (41) must be revised as in (42).

(42) $[_V$ CL – CL – CL – V – SA]

Finally, for some conjugations a tense marker appears between the verb stem and the subject agreement. One example of this is the *-r-* in French indicating future tense; (42) becomes (43).

(43) $[_V$ CL – CL – CL – V – TNS – SA]

It is interesting to note that (43) is the kind of representation often attributed to concatenative languages like Swahili and Chilcotin. Barrett-Keach (1980) suggests (44) for Swahili;[6] cf. also Morrison (1986) for a similar representation in Chilcotin.

(44) $[_V$ NEG + SA + TNS + OA + V + FINAL]

Now we have seen that the clitics in French, Italian, and many other languages serve as licensing elements for pro in argument position. It is now also widely suggested that subject agreement in Italian has the same function through government of [NP,IP] by AGR in INFL. A preliminary conclusion in view of these facts would be to say that inflectional agreement markers (OA, SA) and clitics both have the same function, namely that of licensing pro in argument position. More generally, they are elements sufficient for the syntactic recoverability of null arguments.

Let us use this statement as a working hypothesis since it seems to be a natural, fairly unquestionable, one. Going a step further, one can ask: does the fact that agreement markers and clitics have the same function both in a particular grammar and cross-linguistically mean that they are reflexes of the same syntactic element? In other words: can they be analyzed in the same way? I attempt to answer this question by examining the behavior of "attested" agreement markers with respect to some of the properties of "attested" clitics.

4.2.1. Coordination

Consider first Trentino and Fiorentino, where a subject clitic is obligatory in the second conjunct of a coordinate construction as in (12) above repeated in (45).

(45)a. La canta e la balla.
'She sings and she dances.'
b. *La canta e balla.

Brandi and Cordin (1981) as well as Bouchard (1982) take the contrast in (45) to be evidence that the clitics in Trentino and Fiorentino behave more like agreement markers than real (clitic) pronouns. This conclusion was reached on the basis of the observation that agreement markers are always repeated in the second conjunct, as the French examples in (46) show.

(46)a. Il faut que nous mangions et buvions.
'We must eat and drink.'
b. *Il faut que nous mangions et boire.

Nevertheless, it was argued in the previous section that this interpretation of clitics as agreement markers is valid only under a certain analysis of coordination. Under the approach introduced there, this conclusion is not valid. In other words, the fact that a verbal affix needs to be repeated in the second member of a conjunct is an indication that it can be an agreement marker, but this is not a one to one correlation since certain clitics in certain languages can also be repeated under coordination. There is thus a fuzzy area where agreement markers and clitics intersect with respect to their behavior in coordination.

In conclusion, obligatory repetition under coordination appears to be a necessary but not sufficient condition for an affix to be analyzed as an agreement marker.

4.2.2. Doubling

The doubling phenomenon can be used as another test ground for an eventual distinction between agreement markers and clitics. It is assumed here that doubling can be defined as in (47).

(47) Doubling is a process whereby some grammatical features of a lexical NP in argument position can be reproduced phonologically by an affix on the head of which the NP is an argument.

The two Pied Noir French sentences in (48) involve doubling.

(48)a. L' homme il vient.
the man CL come-3S
'The man comes.'
b. Marie l' aime à Jean.
CL like-3S to
'Marie likes Jean.'

Consider next the English sentence in (49), where the affix *-s* is an agreement marker standing for third person singular.

(49) The man comes.

The characterization of doubling given in (47) applies also to (49) since some features of *the man* are repeated by a suffix on the verb. Obviously we do not want to describe (49) as an instance of doubling similar to the ones studied in chapter 3. One way of getting around this problem is by replacing the word "affix" in (47) by the word "clitic." That would rule out (49) as doubling but would simply bring us back to the original issue of distinguishing between agreement markers and clitics.

A more plausible move would be to stipulate that (47) is an optional process. So, parallel to (50) we find (51) in Colloquial French, but there is no equivalent to (51) in Standard English with regard to (49), as shown in (52).

(50) Jean il vient.

(51)a. Jean vient.
b. Il vient.

(52)a. *The man come.
b. *Comes.

This is also problematic, however, because if one finds a language where the affix representing, say, the subject is never doubled by a lexical argument then one would have to conclude that the affix is an agreement marker and not a clitic. This conclusion is not necessarily the right one even though it could be true. In Standard French, doubling is impossible so that we never find sentences such as those in (53) but only the ones in (54).

(53)a. *Jean il vient.
b. *Marie l'aime (à) Jean.

(54)a. Jean vient.
b. Il vient.
c. Marie aime Jean.
d. Marie l'aime.

Clearly, in Standard French, the underlined elements in (53) and (54) are clitics not agreement markers.

Even the other prediction appears to be wrong in some cases: if an affix is obligatorily doubled by a lexical NP then it is an agreement marker. This holds of the English example in (49) but is not true in Trentino where a lexical NP subject must be doubled by a subject clitic, as demonstrated by the ungrammaticality of (55b).

(55)a. El Mario *el* magna.
b. *El Mario magna.

Thus the doubling phenomenon, however it is defined, does not constitute an area where agreement markers and clitics are clearly distinguished.[7] This does not mean that doubling is a spurious notion but only that the phonological repetition of certain features of a lexical argument constitutes doubling only if the verbal affix can independently be defined as a clitic. This is the case in the doubling examples studied in chapter 3. It also means that (49) above and the Italian sentence in (56) are not examples of doubling.

(56) Giovanni arriva.

To conclude, the possible presence of an NP associated (i.e. coindexed) with an affix appears to be a necessary but not sufficient condition for analyzing the affix as an agreement marker

4.2.3. Extractions

Another test suggests itself. Recall that extraction out of doubling constructions is often blocked for Colloquial French as shown in (57); cf. 3.3.

(57)a. Jean il vient.
b. *Un homme il vient.
c. *Qui il vient.

This seems to provide us with a possible test. In (57b), an indefinite NP is barred from appearing with a coreferent clitic. We have seen in 3.3.2 that this is an instance of a more general restriction on LF extractions which blocks quantified NPs from doubling constructions. On the other hand, this restriction does not apply across-the-board since in some languages, such as Pied Noir French, certain Spanish dialects, Trentino, and others, such NPs can be doubled. Therefore the fact that a restriction on the type of NP that can be associated with a verbal affix representing certain features is present in a language strongly suggests that the affix is a clitic, but that the restriction does not apply does not imply that the affix is not a clitic.

We have argued in 3.3 that *wh*-extraction out of doubling constructions is blocked in all the languages/dialects we have examined. We can thus conclude that if *wh*-extraction of an NP is barred when an associated affix is present on the verb then this affix is probably a clitic.

Hence we now have a tentative test to distinguish between agreement markers and clitics. It must be noted, unfortunately, that this test is rather limited in its applications since not all languages exhibit syntactic extractions like *wh*-movement.[8]

To conclude, the presence of restrictions on the type of NP that can be associated with a given affix seems to constitute a necessary but not sufficient condition for analyzing this affix as a clitic.

4.2.4. Obligatoriness

That an affix be obligatorily present on a verb stem is also a strong indication of its status as an agreement marker. But, here again, there is no one-to-one correlation since clitics can also be obligatory in certain languages/dialects. Trentino and Fiorentino are examples of this sort; cf. 3.3.1.4 and 3.3.1.5.[9]

In conclusion, obligatory presence of an affix is a necessary but not sufficient condition for analyzing the affix as an agreement marker

4.2.5. *Morphology*

Zwicky and Pullum (1983) note that morphological idiosyncracies are more characteristic of affixed words than of clitic groups. As an illustration, consider the French paradigms in (58) where it is clear that the agreement markers vary for the same combination of features in different tenses whereas the clitics remain the same.

(58)	Present	Conditional	Future
1S	je parle	je parlerais	je parlerai
2S	tu parles	tu parlerais	tu parleras
3S	il parle	il parlerait	il parlera
1P	nous parlons	nous parlerions	nous parlerons
2P	vous parlez	vous parleriez	vous parlerez
3P	ils parlent	ils parleraient	ils parleront

Thus, morphological changes constitute a necessary but not sufficient condition for analyzing an affix as an agreement marker.

4.2.6. *Conclusion*

There seem to be differences between clitics and agreement markers even though their respective properties are often similar in certain respects. The question of determining the status of a verbal affix representing certain nominal features of the argument of a head is an important one when this status is not clear. This is the case in less studied languages since this status is most likely to have consequences for the behavior of the arguments with respect to certain syntactic phenomena like doubling, null arguments, extraction, and coordination. The question thus remains open pending further research.

It is nevertheless possible to remark that this fuzzy distinction is to be expected considering that both agreement markers and clitics fill the same function in a number of languages, that of licensing elements for pro in argument position.[10] Also, the differences observed in this section and the ones studied in Zwicky and Pullum (1983) clearly suggest that, although very similar, these two types of verbal affixes are probably distinct.

4.3. HISTORICAL SPECULATIONS

There seems to be a possible correlation between the historical development of certain Romance languages and the unclear distinction between agreement markers and clitics. Although the following discussion is highly speculative, I wish to point out that it is natural under the analyses presented in this book.

There is a widespread intuition underlying some works on clitics and agreement markers in a diachronic perspective according to which, as Lapointe (1980, 234) writes, "person/number markers can profitably be analyzed as fossilized clitics which have been left over as the residue of once productive processes from previous states of the languages and which exhibit various differences in behavior across the languages."

This view combined with the analyses presented here would attribute this evolutionary process to the fact that clitics and agreement markers have the same syntactic function and would seem to provide a reason for their resemblance both at any given synchronic slice of the development of a language and across languages. Nevertheless this hypothesis is not completely without flaws. The problem is a natural one and has to do with the fact that clitics and agreement markers are just parts of the syntactic "inventory" of a language. Their interactions with other phenomena can alter the process just described. To illustrate, let us consider the particular case of French.

The traditional view expressed, for example, in Bourciez (1956, 684) and Ewert (1966) cited in Harris (1978, 111), supposes that a phonological change whereby word-final consonants dropped across the board had the effect of gradually eliminating the flexional distinctions of the verbs in Middle French (circa 1300–1600). The effect of this change was to make necessary the use of subject pronouns which then became clitics. This process would not have happened in Italian, Spanish, and Portuguese. These languages thus retained the situation found in Old French and, previously, Latin, where subject pronouns were not obligatory.

On the other hand, more recent studies have reinterpreted the facts. Harris (1978, 113) suggests that the obligatory use of subject pronouns is to be attributed to syntactic word order. He notes, following others, that Old French passed through a verb-second stage.

Subject pronouns could be used to obey this order, that is SVX, and this pattern would have subsequently generalized. Harris, following Price (1971), even points out that the obligatoriness of subject pronouns seems to have preceded chronologically the phonological process alluded to above. This particular process would have accelerated the mandatory presence of subject pronouns and maybe their cliticization.

The reader is referred to Adams (1986) for a defense of an analysis similar to Harris's but in the Government and Binding Framework.

4.4. SUMMARY

Two rather separate issues were discussed here that were raised by the various analyses of the constructions involving clitics. In 4.1, an account was proposed for further symmetry between subject clitics and object clitics. Whereas object clitics are base-generated on the verbal head of the constituent structure, subject clitics are base-generated as adjoined to AGR in INFL and are subsequently affixed to the verb by a late rule of Finite Verb Raising of the type proposed in Emonds (1978). This lead us to an examination of some of the effects of coordination on FVR.

The second issue, addressed in 4.2, concerned the fact that clitics as they are analyzed here have the same syntactic function as agreement markers with respect to the licensing of pro. The question was then raised of determining whether they are just different realizations of a single syntactic notion. A study of some of the constructions in which they are both involved lead to the conclusion that clitics and agreement markers probably have different status. I pointed out that the issue is raised in the first place because clitics and agreement markers often behave in the same way with respect to some phenomena but the problem of providing syntactic tests that would clearly distinguish the two notions still remains. I have not been able to provide necessary and sufficient conditions for analyzing a given affix as a clitic or as an agreement marker.

Conclusion

To conclude this book, I will simply review the main theoretical proposals I have made and try to put them in the more general perspective of research in linguistic theory. The preceding work rests on the major underlying assumption that empty categories are available in natural languages. Each type of empty category is syntactically licensed in a given way. Traces are licensed by the Empty Category Principle and government theory modulates the occurrences of PRO since PRO appears only in ungoverned positions. In examining the distribution of pro both across languages and within particular grammars, I have opted for a theory specifying minimal conditions on its presence in syntactic structures: pro must be coindexed with and governed by some element which carries the set of features necessary for its licensing. I exempt from this statement the English constructions discussed in the introduction which remain uninvestigated.

The above conditions on the presence of particular types of empty categories in a syntactic structure share at least one characteristic: they all derive from the more general and fundamental Recoverability Principle which was presented in Introduction and is repeated in (1) below.

(1) Recoverability Principle
 An empty category must be licensed.

As we have seen, the development of this principle in linguistic theory has followed the familiar path leading to the postulation of fundamental principles which replace language-particular rules.

A review of previous analyses showed how rich agreement in languages like Italian and Spanish was an instance of a licensing device for pro. Pro in these languages is governed by and coindexed with the set of features represented by the verbal agreement. This verbal agreement is, in a certain sense, sufficient for the licensing of pro. Pro is thus licensed by rich verbal agreement and, therefore, "allowed." This is the way in which the intuition that subject-verb agreement is sufficient to allow for the recoverability of missing subjects is implemented in the grammar.

We then turned to clitics and proposed the working hypothesis that subject clitics and object clitics are two different realizations of a single syntactic element with the same function as AGR. It was argued, following others, that object clitics are licensing elements for pro in the argument position they are related to and that pro is, consequently, coindexed with and governed by a clitic. In most cases, the clitic has the same intrinsic features as agreement markers. The main consequence of this assumption, namely, that subject clitics are also licensing elements for pro in external argument position, was then explored. This hypothesis is supported by the cross-linguistic generalization that the languages with subject clitics studied here never "exhibit" simple tensed sentences in which both a lexical NP subject and a subject clitic are missing. This generalization is the same as the one that can be stated for internal arguments: in languages with object clitics, one never finds a simple sentence where both the object clitic and the lexical NP object are missing (when the verb subcategorizes for an object).

Therefore, the intuition behind this generalization is the same as the one concerning subject-verb agreement, that is, clitics are sufficient devices for the recoverability of missing arguments. The implementation of this intuition follows the same path: clitics are licensing devices for pro in external or internal argument position. When no lexical material is generated in argument position in tensed sentences, this position is occupied by pro which must be licensed by a clitic. The above generalization is thus derived since pro is properly licensed only when a clitic is present.

The other working hypothesis has to do with the cross-linguistic characterization of subject clitics and implies that they are generated in AGR in INFL, where they are in a position to license pro in [NP,IP]. It was argued that this is so in all the Romance languages/dialects exhibiting subject clitics.

In order to account for the fact that clitics and lexical NPs are in complementary distribution in a number of languages, I have assumed, following others, that clitics are Case absorbers. A phonologically realized NP in the argument position coindexed with a clitic is thus non Case-marked and blocked by the Case Filter. Alternatively, it can be assumed that an NP without Case cannot be lexicalized in PF in accordance with Bouchard's (1982) proposals. As far as the behavior of pro with respect to Case is concerned, I tentatively assumed that pro does not need Case.

In terms of learnability and UG, complementary distribution between a clitic and a phonologically realized NP is considered the unmarked case. The parameter involved would thus have the form given in (2).

(2) A clitic absorbs a Case feature of the lexical head to which it is affixed.

The language-learner must first determine that his/her language includes some clitic paradigm(s). It is supposed that positive evidence is available that indicates that certain pronouns (the clitics) behave differently from regular NPs. For each clitic paradigm (subject, object, etc.), the unmarked case is set, namely, that there is complementary distribution between clitics and phonologically realized NPs.

Linguistic theory, as it is understood here, establishes a close relationship between learnability and cross-linguistic variation. The analyses I have suggested are thus embedded in a certain perspective that allows us to propose, not only some account for various syntactic constructions, but also some account of learnability.

In this respect languages are divided into two groups: those with clitics and those without. Further, among those with clitics, various combinations of clitic paradigms are found.

In certain languages there is no complementary distribution between some clitic paradigm and phonologically realized NPs. These languages are said to allow clitic-doubling. The main hypothesis defended here is that subject doubling and object doubling are made possible by the same mechanisms. For an NP coindexed with a clitic to be phonologically realized, Case must not be absorbed by the clitic. In other words, in certain languages clitics need not absorb the Case feature of the head to which they are affixed. Case can thus be assigned in the familiar way to the NP position.

The parameter involved here is once again the one in (2). In the case of each clitic paradigm, the language-learner may set parameter (2) negatively. In other words, a clitic need not absorb the Case feature. Doubling constructions involving each clitic paradigm constitute positive evidence that the parameter in (2) must be set negatively. Therefore, languages with clitics are divided into two groups: those with doubling and those without.

Case-assignment in doubling constructions is the source of a major asymmetry between subject clitics and object clitics. Whereas a dummy Case-assigning device is necessary for object position, this is not so in subject position. The particular theory of Case presented in 3.2.2 may account for this asymmetry.

Even though Case can be assigned in familiar ways in doubling constructions in spite of the presence of clitics, there are, in certain languages, some restrictions on the type of NPs that can be clitic-doubled. We argued that the fact that quantified NPs are often barred from clitic-doubling is to be related to the other restrictions concerning syntactic extractions of clitic-doubled NPs.

A characterization in terms of c-chains of the relationship holding between a clitic and an argument position allowed us to systematically block the presence of a bound NP in the argument position involved in a c-chain, be it A-bound or Ā-bound. This last point is the source of some cross-linguistic variation, which, it was argued, is to be accounted for by the particular syntactic level at which a c-chain is formed. Most languages disallow syntactic extractions out of doubling constructions and some also disallow LF extractions out of this construction. In the former case, c-chains are formed at S-structure, therefore permitting the breaking of the chain at LF where QR applies. In the latter case, c-chains can never be broken with the effect that syntactic extractions like *wh*-movement as well as LF extractions like QR are blocked. C-chains are the same for subject clitics and object clitics.

More generally, I have offered a theory of clitics that unifies the various clitic paradigms found in natural language. They are all considered here to be different realization of the same syntactic element. The differences between the various clitic paradigms must follow, in my analysis, from independently motivated asymmetries and not from intrinsic differences between the different clitics.

I then discussed some related issues, especially concerning subject cliticization where it was proposed that subject clitics in INFL are affixed to the verb by a late rule of Finite Verb Raising.

The different analyses suggested attribute to clitics and agreement markers the same role in the syntactic structure, namely that of licensing pro. I thus examined the question of determining whether clitics and agreement markers are the same elements and concluded, tentatively, that they were not, on the basis of their differential behavior with regard to various constructions and operations.

Finally, it is worth noting that the three conditions on the presence of particular types of empty categories alluded to earlier share another characteristic: they all involve the theory of government. It appears therefore that the Recoverability Principle and the theory of government are intimately related. Since the theory of government seems to play such a central role in the theory of recoverability, it will probably prove to be of primary importance in eventual accounts of the English constructions discussed in the introduction.

The interesting question at this point is: why should government and recoverability be so intimately related? An answer to such a question is difficult to provide since it amounts to asking why government is such an important concept in the theory of grammar. It is nevertheless intriguing, and especially revealing, that syntactic recoverability is further confirmation of the importance of the concept of government in linguistic theory. It would certainly be interesting to attempt a unification of the various licensing operations appealed to for pro, trace, and PRO in order to achieve a more restrictive theory of universal grammar.

Notes

CHAPTER ONE

1 Cf. Baker (1979), who presents arguments that only positive evidence is available to the child.

2 Chomsky (1986b, 4) assumes that only X^{0} (minimal projection) and X^{max} (maximal projection) are "visible" for Move-α; see 1.1.1 concerning the X′ system.

3 The term "external argument" is introduced in Williams (1980). See also Williams (1981) concerning the further distinction between external argument and "internal" argument.

4 Chomsky (1986b) develops an explicit account of barriers in which this notion is built into the definition of government. Projections are not barriers in themselves anymore but only in relation to the position of the governee with respect to the governor. This shift has no direct effect on the analyses presented here.

5 Chomsky (1982) points out that given the functional definition of empty categories he adopts together with a particular definition of variables, condition C of binding theory can be dispensed with. I will have nothing to say here about this possibility. Nevertheless, it remains that even though condition C is not built into the theory, its effects still exist, which is the reason I opt for maintaining it as an explicit part of the theory.

6 This is true only under the assumption that AGR is the nominative Case-assigner; cf. 2.4 for more on nominative Case-assignment.

7 English and Italian, in this respect, are distinguished by the presence in the latter of an empty category in the object position of V in the sentences in (40). In English, the verbs are simply intransitive.

Rizzi (1986b) argues, on the basis of empirical evidence, that the empty object in Italian is "active" syntactically whereas in English there is no empirical evidence for the presence of an empty object. For instance, the empty object in Italian can serve as antecedent for an anaphor, it can be a controller, and it can be the subject of a small clause; cf. Roberge (1989) for a reanalysis of this property in the framework presented in this book and Bouchard (1989) and Williams (1986) for opposing views. Massam and Roberge (1989) discusses the availability of pragmatically well-defined null objects in English.

8 Brody (1984, 363) discusses a similar problem with respect to the English sentence in (i).

(i) Who$_x$ did PRO$_x$ losing his$_x$ way annoy e_x?

In (i), the pronominal anaphor PRO is locally $\bar{A}$-bound by the antecedent (*who*) of *e*. According to the contextual definition of empty categories given in Chomsky (1981, 330), a locally $\bar{A}$-bound empty category is a variable. In other words, a locally $\bar{A}$-bound empty category is a non-pronominal non-anaphor. On the other hand, Brody points out that the first empty category in (i) must be a pronominal because it would otherwise violate the ECP.

CHAPTER TWO

1 The material in 2.1 and 2.2 is presented in a condensed form in Roberge (1985;1986).

2 The main source for the data in these dialects is Brandi and Cordin (1981) and Safir (1985). The sources for the individual examples will not be mentioned.

3 I use the term Colloquial French to refer to the dialect spoken in Québec. It shares some syntactic properties with the French dialects spoken in some provinces of France. Some relevant data for this dialect can be found in Laurendeau, Néron and Fournier (1982), Carroll (1982) and Roberge (1986); cf. also 3.1.2. Pied Noir (or Pataouète) is the French dialect spoken in North Africa, mainly in Algeria. The Pied Noir data in the rest of this book come from a single informant, Guy Nephtali, originally from the Oran region and now living in Québec. A good source of confirmation for these data can be found in Bacri (1983) and in Lanly (1970).

4 Cf. Everett (1985) for some Pirahã data and Huang (1984) for Pashto.

5 But cf. Safir (1985) who treats German as Nom-drop language in which only pleonastic subjects can be missing. Huang (1984) presents evidence that German is not pro-drop but zero-topic. Cf. also Travis (1984) on pleonastic subjects in German.

6 Rouveret (1980, 99) makes a similar statement: "Dans [les langues] qui ont développé des formes de clitiques sujets, l'apparition du clitique est obligatoire en l'absence d'un sujet plein nominal ou pronominal: celles

qui n'ont pas de clitiques sujets admettent des sujets phonétiquement nuls dans les propositions finies."

The hypothesis put forward by Rouveret in order to account for this observation is that the subject clitic is a spell-out of the nominative Case feature and is therefore obligatory if no other lexical element in subject position can receive this Case. This is very similar to Borer's (1981) analysis of clitics in VP, PP and genitival constructions. This hypothesis is incompatible with the one adopted here that treats clitics as licensing elements for pro.

7 A potential counterexample to (6) is the Raising construction in French.

(i) Jean semble être malade.
'Jean seems to be sick.'

This construction is generally represented in D-structure as in (ii),

(ii) *e* semble Jean être malade.

where the NP subject *Jean* of the embedded infinitival clause is forced to move in order to get Case and where the subject position of *sembler* does not get an external Θ-role. I return to this construction in 2.2.1, where it is shown that there can be an expletive pro in subject position of such verbs as *sembler*.

8 This approach is compatible with, though different from, Zubizarreta's (1982b) assumptions regarding the properties of pro.

Cf. Rizzi (1986b) and Adams (1986) for approaches requiring a two way condition on the appearance of pro.

The details of Rizzi's theory are presented in 1.2.3. In his account the licensing domain of pro is that of Case-assignment. Adams's licensing condition is less restricted than Rizzi's since it implies only government by a head. She further argues that government holds in only one direction (to the left or to the right) in any given particular grammar.

9 There is a construction with empty NP in PPs in French that seems to cause problem for such an analysis since French does not have clitics or agreement markers affixed to prepositions as in examples (12) and (13) in the text. The following is based on Zribi-Hertz (1984) who presents arguments against interpreting the sentences in (i) below as instances of "ellipsis" or as the result of movement.

(i)a. Cette valise, je voyage toujours avec *e*.
this suitcase I travel always with
'This suitcase, I always travel with it.'

b. Les arbres, Pierre se cache toujours derrière *e*.
the trees REFL hides always behind
'Trees, Pierre always hides behind them.'

c. Notre-Dame,	vous	venez	de	passer	devant	*e*.
	you	came	of	to + pass	in + front	

'Notre-Dame, you just passed it.'

Rather, Zribi-Hertz assumes that the empty position in (i) is occupied by pro even though, as she notes, the usual conditions on pro regarding its licensing are not met.

It is not obvious that this type of construction should be analyzed in this way. The empty category in (i) could be a variable bound by the topic, similar to the Chinese examples provided in 1.2.2 since it must refer to an item mentioned in a previous discourse as in (ii); cf. Tuller (1986).

(ii) A: Pourriez-vous m'indiquer où se trouve Notre Dame?
'Could you tell me where Notre-Dame is?'
B: Mais vous venez de passer devant *e*!
'You have just passed it!'

Note also that it is different from preposition stranding in English which is not available in French; cf. Kayne (1981b) for a detailed analysis and Vinet (1979) and Bouchard (1982, 235-6) for discussions of a parallel phenomenon in Colloquial French.

(iii)a. Who did you vote for?
b. *Qu'est-ce que nous venons de passer devant?
'What did we just pass in front?'

Another, less attractive option would be to relate the fact discussed here to a similar phenomenon in English. Consider the sentences in (iv).

(iv)a. Can I come with? (film: *Paris, Texas*)
b. Oh good! You brought them with! (film: *Eating Raoul*)

The only similarity between the sentences in (iv) and the ones in (i) is that in both cases the gap follows a preposition. The two differ in that the English case seems to be limited to the preposition *with* and the gap always refers to the second person pronoun, *you*. The phenomenon is not so restricted in French.

I leave the question open pending further research, especially concerning why the construction is limited to object of prepositions as in (i) if a "topic-variable" type of analysis is adopted. Note finally that an interpretation of the gap in (i) as being pro causes problem for other analyses of this empty category such as the ones in Chomsky (1982) and Rizzi (1986b).

10 Possessive NPs in Turkish seem to behave in the same way with respect to pro except that, in this language, pro is in the specifier (or subject) position of NP where it is licensed by an agreement marker on the head noun; cf. Kornfilt (1984).

(i) pro_i ɨstakoz-um_i
lobster-1S
'my lobster'

(ii) pro_i ɨstakoz-$umuz_i$
lobster-1P
'our lobster'

It is not clear here that the head noun assigns a Θ-role to the specifier. Massam (1985), following Chomsky (1986a), assumes that in the examples below *reading the book* and *story* do assign a Θ-role to the specifier.

(iii)a. John's reading the book disturbed me.
b. John's story disturbed me.
(Chomsky 1984)

Cf. also Gruber (1976) and Anderson (1983). Note that if this analysis holds also of the examples in (i) and (ii), it is then predicted that no expletive pro is allowed to appear as the specifier of N′ position in languages with clitics or agreement markers on N. I cannot verify the value of this prediction. As for genitive Case-assignment in this construction see Borer (1984) and chapter 3, footnote 15.

11 That subject-verb agreement is "rich" in European Portuguese is supported by the following paradigm from Raposo (1985). The morphology of the agreement markers is the same for all conjugation classes.

(i) *comer* 'to eat'

1S	eu comer + ϕ I	1P	nós comer + mos we
2S	tu comer + es you	2P	vós comer + des you
3S	ele comer + ϕ he	3P	eles comer + em they

For example, in the following sentence with a lexical NP as the subject of an infinitive verb, the verb must bear the relevant agreement marker.

(ii) Será dificil [os deputados aprovar*em* a propostal].
'It will be difficult for the deputies to approve the proposal.'

12 Some N.It. dialects have gaps in the paradigm of their subject clitics. In Trentino, they are 1s, 1p and 2p; cf. Renzi and Vanelli (1982), Brandi and Cordin (1981) and Rizzi (1986a). Furthermore, the subject clitics in Trentino and Fiorentino are obligatory even when a lexical preverbal NP subject is realized. I return to these issues in 3.1.1.

13 Unless mentioned otherwise, the Arabic data found in the rest of this book is due to a single informant, Jumah Hameed, born in Iraq and now living in Canada.

14 I follow Borer (1984) in assuming that the relationship between the object clitic and the verb is one of affixation (characterized by +) as opposed to adjunction as in (i) found for example in Jaeggli (1986, 107).

(i)

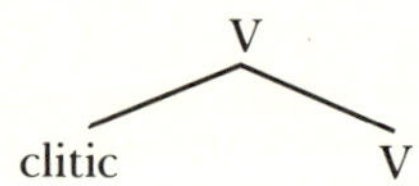

Jaeggli (1985, 2) assumes that the clitic is: "a separate 'word' syntactically. However, as it is dominated by the same level node as the word it is affixed onto, it is considered also to be part of the verb." In this view, the government domain of the clitic is the same as that of the verb. The same is true in Borer's analysis.

15 A similar approach is developed in a different way in Safir (1985).

16 Cf. Kayne (1975) for the various properties of this construction.

17 For analyses involving Verb Raising see Emonds (1978), McA'Nulty (1983), Lobeck and Roberge (1985) and chapter 4.

18 For Rizzi, this holds only of the subject clitics in the N. It. dialects.

19 Note that (54) is the representation adopted in Safir (1985).

20 Cf. Jaeggli (1982;1986), Borer (1984) and 3.2 for more on River Plate Spanish A.

21 Cf. Hurtado (1985), Borer (1984) and 3.2 for more on River Plate Spanish B.

22 But see the introduction for a discussion of some English constructions with no overt subject.

23 This type of analysis also has the non trivial advantage of providing an account for the apparent *that-t* violations; cf. 1.2.1.

24 For a discussion of the *that-t* violations see in particular Zubizarreta (1981;1982a) and Bouchard (1982). For Zubizarreta, they are the result of a strategy similar to the *que/qui* alternation in French allowing extraction of the subject. In Portuguese, such an alternation would exist even though it is not morphologically realized, i.e. there is a unique complementizer *que* standing for the nominal equivalent of French *qui*. For Bouchard, the violations are to be related to an empty resumptive pronoun strategy.

25 The differences in Travis's use of I′, V′, and N′ are not crucial for the point I want to make here.

Borer (1986, 400) suggests the following base-generated structure for free inversion constructions.

(i)

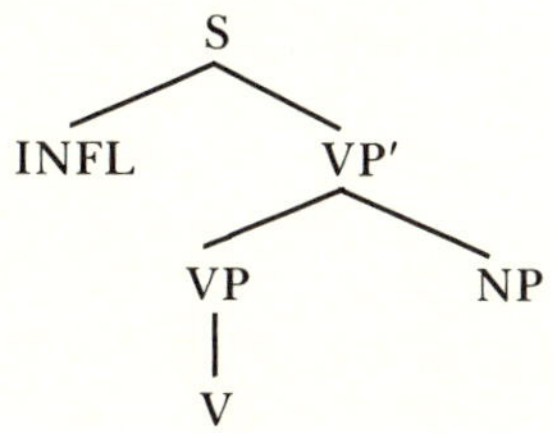

26 As pointed out in Koopman (1984, 212), there is an apparent problem with this claim with regard to inflected infinitives in Portuguese (cf. 2.2.1) in which the subject is marked for nominative Case even though INFL is [−TNS]. Koopman proposes an alternative account of this fact in terms of a V-movement to INFL similar to the one she argues for in Vata and Gbadi. See also Villiard (1985) who argues explicitely for TNS as nominative Case-assigner.

27 But see Culicover and Rochemont (1983) and Pollock (1985) for arguments against analyses of the definiteness effect such as that of Safir's (1985).

28 This neuter clitic *e* becomes *gl'* when followed by a vowel.

(i)	Gl'	ha	portato	la	torta	una	ragazza.
	CL(neuter)	has	brought	the	cake	a	girl

'A girl brought the cake.'

29 It is possible to translate the structures in (88) and Safir's proposal into the Barriers framework. Chomsky (1986b, 14) suggests defining sisterhood in terms of maximal projections instead of lexical projections. In (i), which corresponds to (87) in the text, both NPs are sisters of VP.

(i)

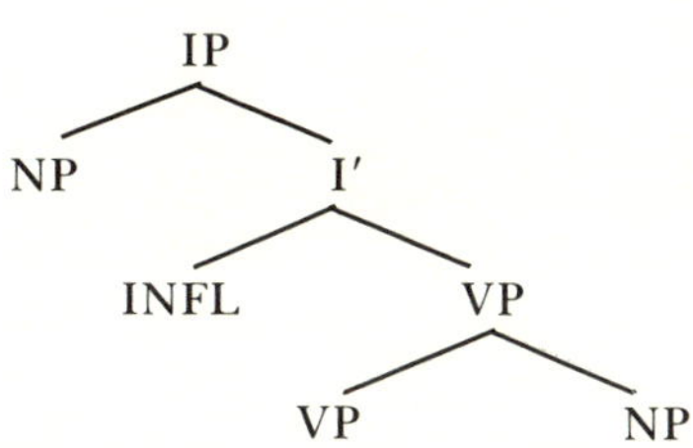

Borer (1986) adopts a structure similar to (88b) in the text but without an [NP,S] position. In her analysis, there is also direct Θ-role assignment to the postverbal NP.

30 Note that this might be a problem for the analysis in 2.2.5 according to which clitics absorb Case since the preverbal impersonal clitic in (80) would not do so because there is no Case to be absorbed. I delay discussion of this to 3.1.4 and 3.2.2 where I argue that clitics do not "have to" absorb Case.

31 This is true also for the free inversion construction involving "unaccusative" verbs in the sense of Perlmutter (1978) or "ergative" in the sense of Burzio (1981) and where the subject is base generated in direct object position.

(i)a. Arriva Giovanni.
'Giovanni arrives.'

b.

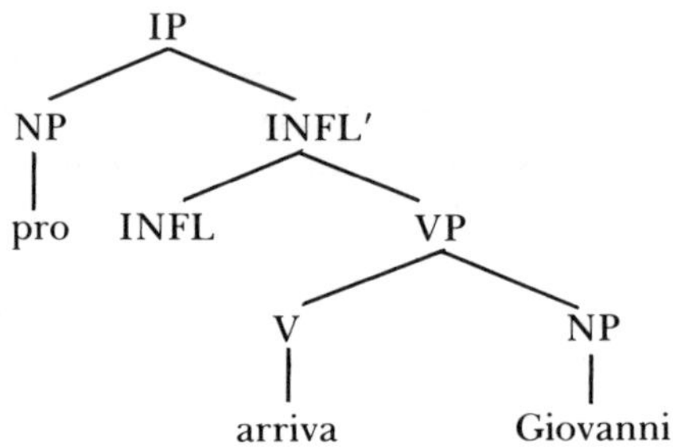

32 The construction with *faire* is often referred to as the "typical" causative construction. In most cases, the properties of the *faire* + infinitive sentence are found with other verbs like *laisser* and some perception verbs such as *voir, entendre, regarder, écouter*; cf. Kayne (1975).

33 The numerical indices in (98) are used for the sake of presentation and have no theoretical significance.

34 Assume here that *à* transmits dative Case from the verb to the NP; cf. 3.2.2.

35 Cf. Goodall (1984, 174) for a discussion of the possible, although marginal, result in (i) based on adjacency requirements on Case-assignment.

(i) ?Marie l_j' a fait la_i manger pro_i pro_j.

36 There is one potential incompatibility between Goodall's analysis and the one presented here. In causative constructions, a very limited number of verbs allow their subject to be either accusative or dative, as shown by the examples in (i).

(i)a. Marie a fait écrire l'enfant.
b. Marie a fait écrire à l'enfant.
'Marie made the child write.'

Goodall (1984, 156) assumes that verbs such as *écrire* have an implicit object which receives the accusative Case; dative Case is assigned to the subject. In (ia), *l'enfant* is assigned accusative Case.

It might be the case that such an implicit accusative object is pro_{arb} in the sense of Rizzi (1986b) in (ib). We would therefore have an example of an argument pro in French that is not licensed by a clitic, in contrast with what I have proposed up to now.

CHAPTER THREE

1 Some of the material in this chapter, especially 3.1 and 3.3, is presented in a condensed and form in Roberge (1986;1988) respectively.

2 There are no subject clitics for 1S, 1P, and 2P in Trentino; cf. (30) in chapter 2, 3.1.4, and Renzi and Vanelli (1982).

3 I use the tripartite expansion of S (=IP) used in Safir (1985) since the differences with the one adopted here are not crucial for our discussion.

4 The data for this dialect come from Laurendeau, Néron and Fournier (1982), Carroll (1982) and my own intuition. Note that the third person pronouns have the following surface forms in this dialect:

3SM:	il	→ [i]
3SF:	elle	→ [a]
3PM:	ils	→ [i]
3PF:	elles	→ [i]

5 But see Nemni (1973), cited in Carroll (1982), who points out that it is difficult to build an argument on the basis of such facts. Nevertheless, Dubuisson, Emirkanian and Lemay (1983, 26) in their acquisition paper on dislocations and topicalizations note: "Une phrase avec reprise du sujet comporte un SN (NP) sujet et un clitique sujet tel qu'illustré en (i). À l'oral elle ne comporte généralement ni accent sur le constituant disloqué, ni pause".

(i)a. Marc il compte jusqu'à dix.
he counts up + to ten
'He counts to ten.'
b. Se bouton il sert à alumé le jeu. (sic)
this button it serves to turn + on the game
'This button is used to start the game.'

This study also points out other interesting facts. The authors observe that what they call the "reprise du sujet" – as in (i) which I have called "subject doubling" here – differs in frequency of use in the written language and in the spoken language. The use of subject doubling in the written language decreases from grade two to grade six whereas it increases in the spoken language during the same period. The proper interpretation of this is open, but it seems reasonable to assume that this is so because the written language learned at school corresponds more closely to the standard dialect.

6 Tellier (1986) suggests an analysis of floating quantifiers in C.Fr. that could be used to argue for the presence of pro in a construction with a subject clitic and without a lexical NP. There exist some floating quantifiers in C.Fr. that appear in sentence final position as in (i).

(i) a. Ils ont parlé au directeur *tout le monde*.
'They all spoke to the director.'
b. Ils ont pas parlé au directeur *personne*.
'None of them spoke to the director.'

Tellier reports that a quantifier like *personne* must bind an empty category at S-structure. Crucially, this empty category must apparently be [+pronominal]. Consequently, the sentences in (ii) are acceptable with pro in argument position whereas the ones in (iii) are barred.

(ii) a. *pro*$_i$ ils$_i$ ont pas parlé à mon père *personne*.
'None of them spoke to my dad.'
b. Je les$_i$ ai pas vus *pro*$_i$ *personne*.
'I saw none of them.'

(iii) a. **Les enfants* ont pas parlé à mon père *personne*.
'The children haven't spoken to my dad.'
b. *J'ai pas vu *les enfants personne*.
'I haven't seen the children.'

Similarly, the sentences in (iv), where *personne* binds PRO are grammatical.

(iv) a. *Les enfants*$_i$ ont pas voulu PRO$_i$ parlé à mon père *personne*.
'None of the children wanted to speak to my dad.'
b. J'ai pas demandé *aux enfants*$_i$ de PRO$_i$ me rencontrer *personne*.
'I didn't ask any of the children to meet me.'

The interesting fact here is that a sentence such as the one in (iia) is excluded when a phonologically realized NP subject is present, i.e. when there is subject doubling.

(v) *$Les\ enfants_i$ ils_i ont pas parlé à mon père *personne.*

This is accounted for if we assume that pro can be bound by *personne* in (iia) and that there simply is no pro in (v) since *les enfants* occupies the [NP,IP] position occupied by pro in (iia).

7 This dialect has not received much attention in the literature except for some scattered remarks such as in Aoun (1979: footnote 10) and Jaeggli (1982, 56, footnote 17); cf. also footnote 8 below. Note the following surface forms of the 2S and 3S/P subject pronouns in this dialect:

2S:	tu	→ [tʃi] or [tʃy]
3SM:	il	→ [i]
3SF:	elle	→ [ɛl]
3PM:	ils	→ [i]
3PS:	elles	→ [ɛl]

8 A point worth noting here is that PN French exhibits both subject doubling and object doubling. It is the only Romance dialect or language that I am aware of that has this property. With respect to object doubling in French, this property seems to be more widespread than the literature on the subject would indicate. It appears that other French dialects from the South West of France also have this construction. Bourciez (1956, 690) writes: "Dialectalement, il s'est introduit au Sud-Ouest, sous l'influence du gascon une habitude qui y est encore très vivante, et qui consiste à faire précéder de *à* le pronom complément principal répété par emphase." He provides the following examples:

(i) Montaigne

Que	je	t_i'	escoute	à	toy_i.
that	I	CL	listen	to	you

'I listen to you.'

(ii) Molière

Comment,	tu	me_i	traites	à	moi_i	avec	cette
how	you	CL	treat	to	me	with	this

hauteur
haughtiness
'What! You treat me with such haughtiness!'

According to Bourciez (1956, 458), this "habit" in Gascon and Bearnais is to be attributed to the influence of Spanish. Lanly (1970, 222) proposes something similar with respect to PN French.

As is the case in River Plate Spanish (cf. Jaeggli (1980;1982)), many factors interact with the grammaticality judgements of object-doubling constructions in PN French. For example, the doubled NP must often be [+animate]. I do not discuss these matters here.

9 According to this definition, α c-commands β if the first maximal projection dominating α also dominates β, and α does not contain β. A maximal projection is the highest projection of X^{o}, X^{N}. In the following structure:

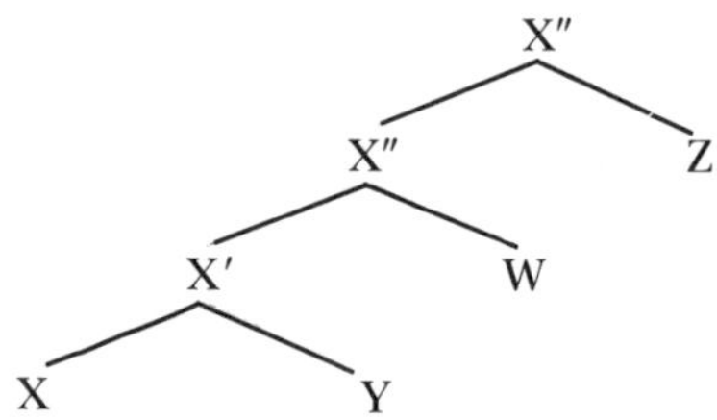

X c-commands Z as well as Y and W; cf. Safir (1985, 6-8).

10 Only the affix status of the clitic is discussed here. Whether the clitic is a suffix, a prefix, or even an infix is not relevant since this is presumably a language specific property of clitics.

11 The Tigre and Lebanese Arabic examples are from Borer (1984, 97) who points out that clitic doubling is not available in PPs in Hebrew.

12 An anonymous reviewer pointed out that Modern Greek and some dialects of Spanish have object-doubling constructions which require no dummy Case-marker. According to M. Guerssel (personnal communication), the situation is the same in Berber. Sportiche (1983, 219) cites Algonkian as a counterexample to Kayne's generalization and suggests that French Complex Inversion illustrated in (i) is another potential counterexample to Kayne's generalization.

(i) Pierre$_i$ viendra-t -il$_i$?
come-FUT-3S-CL
'Will Pierre come?'

This would be true if it can be shown that Complex Inversion is really a construction involving subject doubling; this is not obviously the case. See 3.4 for a discussion.

13 The number of bars allowed for each projection is not relevant here.

14 The same can be said to hold for dative Case-assignment in causative

constructions in French where the preposition *à* is used in a very similar way; cf. footnote 32 in chapter 2.

15 Borer (1984, 46-8) argues that genitive Case in the construct state is assigned by the head noun under specific structural conditions. The genitive Case feature is a potential property of all nouns in Semitic languages, here Hebrew and Arabic.

16 This hypothesis concerning nominative Case-assignment could have interesting consequences for inflected infinitives in Portuguese. Suppose that TNS is always specified in UG as having the [+CA] feature. Then, Case would always be potentially assigned to [NP,IP] in any language, even in infinitival clauses. The difference between English and Portuguese in this respect would be the presence in INFL in Portuguese inflected infinitives – but not in English – of AGR with a [+C] feature. This would account for Case-assignment to [NP,IP] in inflected infinitives in Portuguese and the non-Case-assignment to the same position in English infinitives.

Note however that this theory causes a problem for my suggestions on how to handle free inversion in 2.4 since only the [+CA] feature would be in VP at S-structure where Case is assigned or checked.

17 Jaeggli's (1986, 37) hypothesis analyzing dummy Case-markers as "Case bridges" does not provide a principled explanation for the difference between subject doubling and object doubling. His account is based on the formation of Case pairs and the usual requirement that a single Case be assigned to at most one nominal element. This requirement is formulated as in (i).

(i) Let α be a Case feature.
If there are two or more Case pairs:
(α,β) (α,δ)
then β and δ cannot both be nominal elements.
(Jaeggli 1985, 24)

Clitics are considered nominal elements whereas dummy Case-markers (prepositions) are naturally not nominal elements. Thus, in (ii), an example of object doubling in River Plate Spanish,

(ii) Lo_i vimos a $Juan_i$.
CL saw-1P to
'We saw Juan.'

three Case pairs are formed:
a. (Case feature of the verb, clitic *lo*)
b. (Case feature of the verb, preposition *a*)
c. (Case feature of the preposition *a*, object NP *Juan*)

The Case feature of the verb is associated with only one nominal element, the clitic, and the Case feature of the preposition is associated with only one nominal element, the object NP.

Now in an example of subject doubling like the Colloquial French one in (iii),

(iii) Jean$_i$ il$_i$ vient.
CL come-3S
'Jean comes.'

two Case pairs would be formed:

a. (Case feature of INFL, clitic *il*)
b. (Case feature of INFL, subject NP *Jean*)

This situation is ruled out by (i) since one Case feature is associated with two nominal elements, the clitic and the subject NP. One way around this problem would be to assume that subject clitics are not nominal elements. This would go against our basic assumption that subject clitics and object clitics are not intrinsically different.

18 Cf. May (1977) and Hornstein (1984). I simply assume that this is true of all the NPs in (88) without further comment. Note, however, that the striking similarity in the behavior of indefinite NPs and regular quantified NPs can be used as an argument in favor of analyzing them as quantifiers undergoing QR at LF. The same remark holds of Focus Raising in the sense of Rochemont (1986) (see below); cf. Heim (1982) for a different approach.

19 The sentence in (93c) contains a definite universal quantifier like the one found in the River Plate Spanish example in (88a). This is excluded and should not be confused with similar sentences containing a collective NP, which is not a quantifier as in (i).

(i) Tout le monde$_i$ il$_i$ était là.
all the people CL was there
'Everybody was there.'

The NP in (i) belongs to the class of NPs containing other collective NPs which are undoubtedly not quantifiers such as *famille, classe, foule,* etc.

(ii) a. La famille$_i$ elle$_i$ est venue au party.
'The family came to the party.'
b. La classe$_i$ elle$_i$ est devenue incontrolable.
'The class became uncontrolable.'
c. La foule$_i$ elle$_i$ criait tous le temps.
'The crowd was shouting all the time.'

However, the NP in (93c) is a quantifier similar to the one in (iii).

(iii)	Tout	homme$_i$	(*il$_i$)	aime	travailler.
	all	man	CL	likes	to + work
	'Any man likes working.'				

Subject doubling is possible with collective NPs, (i) and (ii), but impossible with quantifiers, (93) and (iii). Note that collective NPs are usually grammatically singular even though they are semantically plural. Compare (93c) to (i). Cf. also Laurendeau, Néron and Fournier (1982).

20 For an account of the *que* appearing optionally along with a fronted *wh*-word in questions in other French dialects, cf. Lefebvre (1982), Roberge (1983), and Kayne (1984, 231).

21 Jaeggli (1982, 49) provides the example in (i) to argue that a clitic-doubled NP cannot be focused in RP Spanish.

(i) *Yo lo$_i$ vi a JUAN$_i$.
'I saw JUAN.'

This is expected since LF extractions are excluded from object-doubling constructions in this dialect.

22 Cf. 2.4 and footnote 27 in chapter 2 for more on free inversion and the impersonal (or neuter) clitic.

23 Borer (1984) provides independent evidence that the clitic in (111b) cannot be analyzed as a resumptive pronoun; cf. also Comorovski (1986) where the same conclusion is reached.

24 Note however that *(pe) cine* is excluded from relative clauses whether there is a clitic or not; cf. Dobrovie-Sorin (1985, 6).

25 Note however that it is not necessarily true that the possibility of extracting an element out of clitic-doubling constructions entails the availability of clitic-doubling without extracted element. One could imagine a language that allowed only clitic-doubled variables.

Modern Hebrew appears to provide examples of this sort. According to Borer (1984), Modern Hebrew allows clitic-doubling only in the construct state as in (i) because the preposition *šel* can be inserted in order to Case-mark the complement noun.

(i)	beit -o$_i$	šel	ha -more$_i$
	house-his	of	the-teacher
	'the teacher's house'		

This preposition cannot be inserted in PPs and it follows that the complement of a preposition must be empty when a clitic appears on that preposition.

(ii) a. 'it -o$_i$ [e]$_i$
with-him
b. *'it-o$_i$ Dan$_i$
'with Dan'

Extractions out of doubling constructions are restricted in Modern Hebrew. They are possible only in free relatives, not in questions. The interesting point is that a *wh*-word can be extracted from the construct state with a clitic as well as from a PP with a clitic as shown in (iii) and (iv).

(iii) kaniti ma$_i$ še -xašavt 'al -av$_i$ [e]$_i$
bought-I what that thought-you about-it
'I bought whatever you thought about.'

(iv) 'anaxou 'ozrim le-kol mi$_i$ še -beit -o$_i$ [e]$_i$
we help to every who that-house-his
nisraf
burned
'We help everyone whose house burned.'

Although (iib) is ungrammatical, the sentence in (iii) is grammatical with a clitic-doubled variable.

26 A fourth possible analysis suggests itself. It could be suggested that the impossibility of extracting out of doubling constructions, i.e. the ungrammaticality of (i), for instance,

(i) *[$_{CP}$ *wh*$_i$ [$_{IP}$ t_i CL$_i$ + V]]

is due to a violation of the Bijection Principle (Koopman and Sportiche 1981) where the *wh*-operator binds two variables or, alternatively, that two operators (*wh* and clitic) bind one variable. However, for independent reasons, the position occupied by the clitic can be neither an A- nor an $\bar{A}$-position, hence the clitic is neither a variable nor an $\bar{A}$-binder; cf. Borer (1984) and 2.2.5.

27 α properly governs β if and only if α governs β and α is lexical (Chomsky 1981, 273).

28 Cf. also Jaeggli (1986) for an extension of this analysis to extractions out of object doubling constructions in certain Spanish dialects.

29 O. Jaeggli has pointed out to me that his line of research follows a path similar to mine with respect to (137). The differences between the two approaches follow crucially from different conceptions of empty categories; cf. 1.3.

30 This differs from the analysis in Roberge (1988) where a further con-

dition on the nature of what can occupy the argument position is added to (137). The present analysis is simpler and covers the same data.

31 Cf. Hurtado (1984a;1984b;1985) and Jaeggli (1986) for alternative analyses in which the quantified expressions in (142), i.e. the extracted NPs in my analysis, do not bind variables in IP. Their arguments are based on the fact that the sentences in (142) do not exhibit weak crossover effects whereas the same sentences without clitics do.

32 Complex Inversion without a lexical NP is also called Subject Clitic Inversion (Subj-Cl-Inv) in Kayne (1972;1975) who argues that they are the same construction.

33 Kayne (1984) also suggests an analysis of Complex Inversion where not only the lexical NP is found in a position external to IP but also the inflected verb and its affixed clitics; cf. 2.2.4.

34 That is because the Case Filter is derived from the Visibility Hypothesis in Chomsky (1981), which is a condition on Θ-role assignment to A-chains. It is therefore not relevant for NPs in $\bar{A}$-position; see footnote 10.

35 In Colloquial French yes/no questions, the interrogative particle *tu* appears to the right of the inflected verb.

(i) a. Pierre va-tu venir?
b. Marie va-tu chanter?

Contrary to what is the case with Complex Inversion in Standard French, this particle is barred from *wh*-questions.

(ii) a. *Quand Pierre va-tu venir?
b. *Quand Marie va-tu chanter?

It can also appear together with a preverbal referential subject clitic contrary to Complex Inversion as shown in (iii),

(iii) a. Il va-tu venir?
b. Elle va-tu chanter?

and with both a lexical preverbal NP and a subject clitic, which is of course excluded in Standard French.

(iv) a. Pierre il va-tu venir?
b. Marie elle va-tu venir?

36 I will have nothing to say, for example, about middle *se* as in the following examples from Kayne (1975, 395). Cf. Aoun (1985, 173) and Manzini (1983) among others.

(i)a. Cela se dit surtout pour ennuyer les gens.
'That is said especially to annoy people.'
b. Les noisettes se mangent bien en parlant.
'Hazelnuts eat well while talking.'
c. Un tel jouet pourrait s'acheter pour soi-même.
'A toy like that could be bought for oneself.'

37 It is tempting to relate this phenomenon to the ethical dative clitics discussed in Jaeggli (1986, 25) for Spanish. The examples are from Jaeggli (1986, 11); cf. also Lapointe (1980, 257).

(i)a. Este chico no *me* come!
'This kid won't eat (for) me.'
b. No te *me* pongas a gritar así, por favor.
'Don't start screaming like that (on me), please.'
c. *Te* lo llamo al doctor?
'Should I call the doctor for you?'

38 Other verbs with inherent *se* include *s'accroupir* 'to squat,' *s'acharner* 'to persist,' *se blottir* 'to crouch,' *se cotiser* 'to club together,' *se dédire* 'to retract,' *s'ébattre* 'to frolic,' *s'entraider* 'to help one another,' *se fier* 'to rely,' *s'ingénier* 'to contrive,' *s'insurger* 'to revolt,' *se lamenter* 'to lament,' *se méfier* 'to be watchful,' *se méprendre* 'to be mistaken,' *se pavaner* 'to strut,' *se réfugier* 'to take refuge,' *se spécialiser* 'to specialize,' and *se targuer* 'to pride oneself.'

CHAPTER FOUR

1 Note that in the instances of free inversion in Trentino and Fiorentino, the TNS element of INFL is moved in VP prior to S-structure. This does not constitute a problem for FVR. The verb, together with TNS now attached to its right, moves to AGR in PF as in (i).

(i)

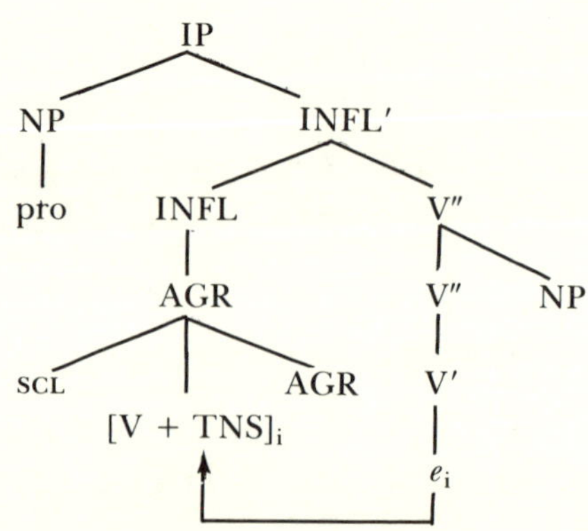

2 Cf. Zagona (1982) for a different analysis of VP anaphora.

3 Cf. also Brandi and Cordin (1981, 2), and Bouchard (1982, 407) for similar views.

4 This is also the case in PN French. Cf. footnotes 4 and 6 in chapter 3 for the surface forms of the third person pronouns in C. French and PN French.

5 The Chilcotin data, an Athabaskan language spoken in British Columbia is due to Morrison (1986).

6 Note further that Barrett-Keach (1986) argues for the structure in (ii) and against the one in (i), where SP indicates subject pronoun clitic and OP indicates object pronoun clitic.

(i)

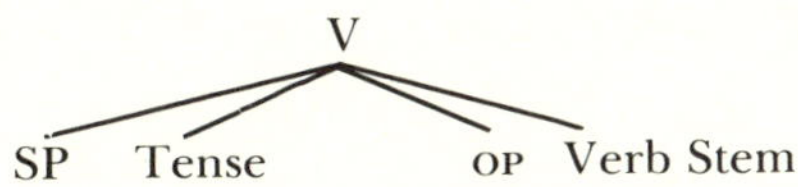

(ii)

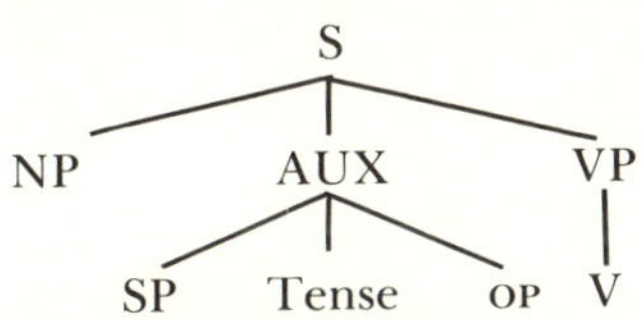

7 Note that Case-absoption cannot be used as a test for the desired distinction since its ultimate effect is the absence of doubling which is not a property specific to agreement markers.

8 Cf. for example Mchombo's (1984) study on object markers in Bantu where he argues that they are clitics and not agreement markers. There seems to be *wh*-extraction in Bantu even when the underlined clitic is present as in (i).

(i) [Kodi ndi nkhuku] ziti zimene mwana uyu ana*zi*ba.
'[Which chickens] did this child steal them?'
(Mchombo 1984:11)

But Mchombo and Mtenje (1983) have argued for the base-generation of interrogative constructions without appeal to movement. Our criterion, even though it is not violated by (i), cannot be used to argue for the clitic status of *zi* in this sentence. Cf. also Bresnan and Mchombo (1985) for a detailed analysis in a Lexical-Functional Grammar framework.

9 Zwicky and Pullum (1983, 503ff) suggest that "arbitrary gaps in the set of combinations are more characteristic of affixed words than clitic

groups." Unfortunately, there is no one-to-one correlation between arbitrary gaps and agreement markers; cf. 3.1.4.

10 See Jelinek (1985) for an entirely different approach where it is suggested that clitics and agreement markers be analyzed as the elements fulfilling the requirements of the Projection Principle. More precisely, they are the true arguments of the verb. For arguments against this type of approach see 1.2.

Bibliography

Adams, M. (1986) "From Old French to the Theory of Prodrop," ms. University of California, Los Angeles.

Aissen, J. (1974) "Verb Raising," *Linguistic Inquiry* 5: 325–66.

– (1979) *The Syntax of Causative Constructions*. New York: Garland.

– and D.M. Perlmutter (1976) "Clause Reduction in Spanish," *Proceedings of the Second Annual Meeting of the Berkeley Linguistics Society*.

Akmajian, A., R.A. Demers and R.M. Harnish (1984) *Linguistics: An Introduction to Language and Communication*. Cambridge: MIT Press.

Anderson, M. (1983) "Prenominal Genitive NPs," ms. University of Connecticut, Storrs.

Anderson, S. and P Kiparsky, eds. (1973) *A Festschrift for Morris Halle*. New York: Holt, Rinehart and Winston.

Aoun, J. (1979) "On Government, Case-Marking and Clitic Placement," ms. Massachusetts Institute of Technology, Cambridge.

– (1981) *The Formal Nature of Anaphoric Relations*. Ph.D diss., Massachusetts Institute of Technology, Cambridge.

– (1985) *A Grammar of Anaphora*. Cambridge: MIT Press.

– and D. Sportiche (1983) "On the Formal Theory of Government," *The Linguistic Review* 2: 211–36.

Bacri, R. (1983) *Trésors des racines pataouètes*. Paris: Belin.

Baker, C.L. (1979) "Syntactic Theory and the Projection Problem," *Linguistic Inquiry* 10: 533–81.

Barrett-Keach, C.N. (1980) *The Syntax and Interpretation of the Relative Clause Construction in Swahili*. Ph.D. diss., University of Massachusetts, Amherst.

– (1986) "Word-Internal Evidence from Swahili for Aux/Infl," *Linguistic Inquiry* 17: 559–64.

Belletti, A. (1985) "Unaccusatives as Case-assigners," ms. Massachusetts Institute of Technology, Cambridge.

– L. Brandi and L. Rizzi, eds. (1981) *Theory of Markedness in Generative Gram-*

mar. Proceedings of the 1979 GLOW Conference. Pisa: Scuola Normale Superiore.
– and L. Rizzi (1981) "The Syntax of *ne*: Some Theoretical Implications," *The Linguistic Review* 1: 117–54.
Birdsong, D. and J.-P. Montreuil, eds. (1988) *Advances in Romance Linguistics.* Dordrecht: Foris.
Borer, H. (1984) *Parametric Syntax: Case Studies in Semitic and Romance Languages.* Dordrecht: Foris.
– (1986) "I-Subjects," *Linguistic Inquiry* 17: 375–416.
– ed. (1986) *Syntax and Semantics.* Vol. 19. New York: Academic Press.
Bouchard, D. (1982) *On the Content of Empty Categories.* Ph.D. diss., Massachusetts Institute of Technology, Cambridge.
– (1989) "Null Objects and the Theory of Empty Categories," ms. Université du Québec, Montréal.
Bourciez, E. (1956) *Éléments de linguistique romane.* Paris: Klincksieck.
Brandi, L. and P. Cordin (1981) "On Clitics and Inflection in Some Italian Dialects," ms. Scuola Normale Superiore, Pisa.
Bresnan, J. and S.A. Mchombo (1985) "Topic, Pronoun, and Agreement in Chichewa," ms. Stanford University.
Brody, M. (1984) "On Contextual Definitions and the Role of Chains," *Linguistic Inquiry* 15: 355–80.
Brua, E. (1938) *Fables bônoises.* Alger: Typo-Litho.
Burzio, L. (1981) *Intransitive Verbs and Italian Auxiliaries.* Ph.D. diss., Massachusetts Institute of Technology, Cambridge.
Carroll, S. (1982) "Redoublement et dislocations en français populaire," in C. Lefebvre ed., tome 1: 291–357.
Cassagrande, J. and B. Saciuk, eds. (1972) *Generative Studies in Romance Languages.* Rowley: Newbury House.
Chao, W. (1981) "Pro-Drop Languages and Non-Obligatory Control," ms. University of Massachusetts, Amherst.
Chomsky, N. (1964) *Current Issues in Linguistic Theory.* The Hague: Mouton.
– (1965) *Aspects of the Theory of Syntax.* Cambridge: MIT Press.
– (1972) *Studies on Semantics in Generative Grammar.* Paris: Mouton.
– (1973) "Conditions on Transformations," in S. Anderson and P. Kiparsky, eds.
– (1975) *Reflections on Language.* New York: Pantheon.
– (1976) "Conditions on Rules of Grammar," *Linguistic Analysis* 2: 330–51.
– (1977) "On Wh Movement," in P. Culicover, T. Wasow and A. Akmajian, eds.
– (1980a) "On Binding," *Linguistic Inquiry* 11: 1–46.
– (1980b) "On Cognitive Structures and Their Development: A Reply to Piaget," in M. Piattelli-Palmarini, ed..
– (1980c) "The Linguistic Approach," in M. Piattelli-Palmarini, ed.

– (1981) *Lectures on Government and Binding*. Foris: Dordrecht.
– (1982) *Some Concepts and Consequences of the Theory of Government and Binding*. Cambridge: MIT Press.
– (1986a) *Knowledge of Language: its Nature, Origin and Use*. New York: Praeger Publishers.
– (1986b) *Barriers*. Cambridge: MIT Press.
– (1989) "Some Notes on Economy of Derivation and Representation," in I. Laka and A. Mahajan, eds.
– and H. Lasnik (1977) "Filters and Control," *Linguistic Inquiry* 8: 425–504.
Comorovski, I. (1986) "Multiple WH Movement in Romanian," *Linguistic Inquiry* 17: 171–7.
Culicover, P. and M.S. Rochemont (1983) "Stress and Focus in English," *Language* 59: 123–65.
– T. Wasow and A. Akmajian, eds. (1977) *Formal Syntax*. New York: Academic Press.
Davis, H. (1987) *The Acquisition of the English Auxiliary System and Its Relationship to Linguistic Theory*. Ph.D. diss. University of British Columbia, Vancouver.
di Sciullo, A.-M. (1982) "Opérateurs et liage, cl, AGR," *Revue de l'association québécoise de linguistique* 2, no. 2: 47–56.
Dobrovie-Sorin, C. (1985) "Clitic Doubling and Relative Clause Formation in Romanian and Spanish," ms. CNRS, Paris.
Dubuisson, C., L. Emirkanian, and L. Lemay (1983) "La dislocation, la topicalisation et la non-reprise du sujet sont-elles des critères de maturation syntaxique?" *Revue de l'association québécoise de linguistique* 2, no. 3: 23–33.
Emonds, J. (1978) "The Verbal Complex V'-V in French," *Linguistic Inquiry* 9: 151–75.
Everett, D.L. (1985) "Pirahã Clitic-Doubling and the Parametrization of Nominal Clitics," ms. Massachusetts Institute of Technology, Cambridge.
Ewert, A. (1966) *The French Language*. London: Faber.
Farkas, D. (1978) "Direct and Indirect Object Reduplication in Romanian," *Papers from the Fourteenth Regional Meeting of the Chicago Linguistic Society*.
Fodor, J. (1980) "Fixation of Belief and Concept Acquisition," in M. Piattelli-Palmarini, ed.
Galambos, S.J. (1980) "A Clarification of the Notion Topic: Evidence from Popular Spoken French," in J. Kreiman and A.E. Ojeda, eds.
Goodall, G.T. (1984) *Parallel Structures in Syntax*. Ph.D. diss., University of California, San Diego.
Gross, M. (1968) *Grammaire transformationnelle du français: Syntaxe du verbe*. Paris: Larousse.
Gruber, J. (1976) *Studies in Lexical Relations*. Amsterdam: North Holland.
Guiraud, P. (1973) *Le français populaire*. Paris: Presses Universitaires de France.

Harris, M. (1978) *The Evolution of French Syntax: A Comparative Approach.* New York: Longman.

Heim, I.R. (1982) *The Semantics of Definite and Indefinite Noun Phrases.* Ph.D. diss., University of Massachusetts, Amherst.

Hornstein, N. (1984) *Logic as Grammar.* Cambridge: MIT Press.

– and D. Lightfoot, eds. (1981) *Explanation in Linguistics.* New York: Longman.

Huang, C.-T.J. (1982) *Logical Relations in Chinese and the Theory of Grammar.* Ph.D. diss., Massachusetts Institute of Technology, Cambridge.

– (1984) "On the Distribution and Reference of Empty Pronouns," *Linguistic Inquiry* 15: 531–74.

Hurtado, A. (1984a) "The Unagreement Hypothesis," in L. King and C. Maley, eds.

– (1984b) "On the Properties of LF′," *Cornell Working Papers in Linguistics* 5.

– (1985) "Clitic Chains," ms. Simon Fraser University, Vancouver.

Jackendoff, R. (1977) *X′ Syntax: A Study of Phrase Structure.* Cambridge: MIT Press.

Jaeggli, O.A. (1980) *On Some Phonologically Null Elements in Syntax.* Ph. D. diss., Massachusetts Institute of Technology, Cambridge.

– (1981) "A Modular Approach to Passives," *Coyote Working Papers in Linguistics* 2: 41–59.

– (1982) *Topics in Romance Syntax.* Dordrecht: Foris.

– (1984) "Subject Extraction and the Null Subject Parameter," *Proceedings of the Fourteenth Annual Meeting of the North Eastern Linguistic Society.* Department of Linguistics, University of Massachusetts, Amherst.

– (1986) "Three Issues in the Theory of Clitics: Case, Doubled NPs, and Extraction," in H. Borer, ed.

– and C. Silva-Corvalan, eds. (1986) *Studies in Romance Linguistics.* Dordrecht: Foris.

Jake, J. (1980) "Object Verb Agreement in Tigre," *Studies in the Linguistic Sciences,* University of Illinois, Urbana.

Jelinek, E. (1985) "The Projection Principle and the Argument Type Parameter." Paper presented at the Winter Meeting of the Linguistic Society of America, Seattle, Washington.

Katz, J.J. (1966) *The Philosophy of Language.* New York: Harper Row.

Kayne, R.S. (1972) "Subject Inversion in French Interrogatives," in J. Casagrande and B. Saciuk, eds.

– (1975) *French Syntax: the Transformational Cycle.* Cambridge: MIT Press.

– (1981a) "Two Notes on the NIC," in A. Belletti, L. Brandi and L. Rizzi, eds.

– (1981b) "On Certain Differences Between French and English," *Linguistic Inquiry* 12: 349–71.

– (1984) *Connectedness and Binary Branching*. Dordrecht: Foris.

– and J.Y. Pollock (1978) "Stylistic Inversion, Successive Cyclicity, and Move NP in French," *Linguistic Inquiry* 9: 595–621.

Keenan, E.O. and B. Schieffelin (1976) "Foregrounding Referents: A Reconsideration of Left-Dislocation in Discourse," *Proceedings of the Second Annual Meeting of the Berkeley Linguistics Society*.

King, L. and C. Maley, eds. (1984) *Selected Papers from the XIII Linguistic Symposium on Romance Languages*. Amsterdam: John Benjamins.

Koopman, H. (1984) *The Syntax of Verbs*. Dordrecht: Foris.

– and D. Sportiche (1982) "Variables and the Bijection Principle," *The Linguistic Review* 2: 139–60.

Kornfilt, J. (1984) *Case Marking, Agreement, and Empty Categories in Turkish*. Ph.D. diss., Harvard University, Cambridge.

Kreiman, J. and A.E. Ojeda, eds. (1980) *Papers from the Parasession on Pronouns and Anaphora*. Chicago Linguistic Society.

Laka, I. and A. Mahajan, eds. (1989) *Functional Heads and Clause Structure*. MIT Working Papers in Linguistics, vol. 10.

Lanly, A. (1970) *Le français d'Afrique du Nord. Étude linguistique*. Paris: Bordas.

Lapointe, S.G. (1980) *A Theory of Grammatical Agreement*. Ph.D. diss., University of Massachusetts, Amherst.

Laurendeau, P., M. Néron and R. Fournier (1982) "Contraintes sur l'emploi du PRO-écho sujet en français du Québec", *Revue de l'association québécoise de linguistique* 1, no. 1–2: 115–28.

Lefebvre, C. (1982) "Qui qui vient? ou Qui vient?: voilà la question," in C. Lefebvre, ed., tome 1: 47–101.

– ed. (1982) *La syntaxe comparée du français standard et populaire: approches formelle et fonctionnelle*. 2 tomes. Office de la langue française, Éditeur officiel du Québec, Québec.

Levin, J. and D. Massam (1985) "Surface Ergativity: Case/Theta Relations Reexamined," *Proceedings of the Fifteenth Annual Meeting of the North Eastern Linguistic Society*, Department of Linguistics, University of Massachusetts, Amherst.

Liejiong, X. (1986) "Free Empty Category," *Linguistic Inquiry* 17: 75–93.

Lightfoot, D. (1982) *The Language Lottery: Toward a Biology of Grammar*. Cambridge: MIT Press.

Lobeck, A. (1985) "Why There Is No VP Anaphora in French." Paper presented at the Winter Meeting of the Linguistic Society of America, Seattle, Washington.

– (1986) *Syntactic Constraints on VP Ellipsis*. Ph.D. diss., University of Washington, Seattle.

– and Y. Roberge (1985) "Coordination, Verb Raising and Subject Clitics." Paper presented at the Winter Meeting of the Linguistic Society of America, Seattle, Washington.

Manzini, M.R. (1983) *Restructuring and Reanalysis*. Ph.D. diss., Massachusetts Institute of Technology, Cambridge.

Massam, D. (1985) *Case Theory and the Projection Principle*. Ph.D. diss., Massachusetts Institute of Technology, Cambridge.

– and Y. Roberge (1989) "Recipe Context Null Objects in English," *Linguistic Inquiry* 20: 134–9.

May, R. (1977) *The Grammar of Quantification*. Distributed by the Indiana University Linguistics Club.

McA'Nulty, J. (1983) "Les clitiques et la c-commande: trois raisons pour appliquer la montée du verbe fini du côté gauche de la grammaire," *Revue québécoise de linguistique* 12, no. 2: 137–51.

McCloskey, J. and K. Hale (1984) "On the Syntax of Person-Number Inflection in Modern Irish," *Natural Language and Linguistic Theory* 1: 487–533.

Mchombo, S.A. (1984) "The Nonexistence of Verb-Object Agreement in Bantu," ms. University of Malawi and Massachusetts Institute of Technology, Cambridge.

– and A.D. Mtenje (1983) "Noncyclic Grammar," *Linguistic Analysis* 11: 219–36.

Meyer-Lübke, W. (1899) *Grammatik der romanischen Sprache, vol III: Romanische Syntax*. Leipzig.

Morrison, C. (1986) *Lexical Phonology of Chilcotin*. Master's thesis, University of British Columbia, Vancouver.

Neidle, C. and R. Nuñez Cedeño, eds. (1987) *Studies in Romance Languages*. Dordrecht: Foris.

Nemni, M. (1973) *Vers une définition phonétique et syntaxique de l'incise en franco-canadien et en français standard*. Ph.D. diss., University of Toronto, Toronto.

Perlmutter, D. (1971) *Deep and Surface Structure Constraints in Syntax*. New York: Holt, Rinehart and Winston.

– (1976) "Evidence for Subject Down-Grading," in J. Schmidt-Radefeldt, ed.

– (1978) "Impersonal Passives and the Unaccusative Hypothesis," *Proceedings of the Fourth Annual Meeting of the Berkeley Linguistics Society*.

Pesetsky, D. (1982) *Paths and Categories*. Ph.D. diss., Massachusetts Institute of Technology, Cambridge.

Piattelli-Palmarini, M., ed. (1980) *Language and Learning. The Debate Between Jean Piaget and Noam Chomsky*. Cambridge: Harvard University Press.

Picallo, M.C. (1985) *Opaque Domains*. Ph.D. diss., The City University of New York, New York.

Pollock, J.-Y. (1985) "Il," ms. Université de Paris 12, Paris.

– (1988) "Verb Movement, UG and the Structure of IP," ms. Université de Haute Bretagne, Rennes.

Price, G. (1971) *The French Language: Present and Past*. London: Edward Arnold.

Quicoli, C. (1980) "Clitic Movement in French Causatives," *Linguistic Analysis* 6, no. 2: 131–86.

Raposo, E. (1985) "Case Theory and INFL-to-COMP. A Study of the Inflected Infinitive in European Portuguese," ms. University of California, Santa Barbara.

– (1987) "Romance Infinitival Clauses and Case Theory," in C. Neidle and R. Nuñez Cedeño, eds.

Renzi, L. and L. Vanelli (1982) "I pronomi soggetto in alcune varieta romanze," in *Studi in onore di G.B. Pellegrini*, Padova.

Rivas, A. (1977) *A Theory of Clitics*. Ph.D. diss., Massachusetts Institute of Technology, Cambridge.

Rizzi, L. (1982) *Issues in Italian Syntax*. Dordrecht: Foris.

– (1986a) "On the Status of Subject Clitics in Romance," in O.A. Jaeggli and C. Silva-Corvalan, eds.

– (1986b) "Null Objects in Italian and the Theory of pro," *Linguistic Inquiry* 17: 501–57.

Roberge, Y. (1983) "Les interrogatives sans inversion: gouvernement du noeud INFL et hypothèse d'identité structurale", *Revue de l'association québécoise de linguistique* 3, no. 2: 119–31.

– (1985) "On Doubling and Null Argument Languages," *Proceedings of the Sixteenth Annual Meeting of the North Eastern Linguistic Society*, Department of Linguistics, University of Massachusetts, Amherst.

– (1986) "Subject Doubling, Free Inversion, and Null Argument Languages," *Canadian Journal of Linguistics* 31: 55–79.

– (1988) "Clitic-Chains and the Definiteness Requirement in Doubling Constructions," in D. Birdsong and J.-P. Montreuil, eds.

– (1989) "On the Recoverability of Null Objects," in *Selected Papers from the Eighteenth Linguistic Symposium on Romance Languages*. Amsterdam: John Benjamins.

Rochemont, M.S. (1986) *Focus in Generative Grammar*. Amsterdam: John Benjamins.

Roeper, T. and E. Williams, eds. (1987) *Parameter Setting*. Dordrecht: Reidel.

Ross, J. (1967) *Constraints on Variables in Syntax*. Ph.D. diss., Massachusetts Institute of Technology, Cambridge.

Rothstein, S. (1983) *The Syntactic Forms of Predication*. Ph.D. diss., Massachusetts Institute of Technology, Cambridge.

Rouveret, A. (1980) "Sur la notion de proposition finie, gouvernement et inversion," *Langages* 60: 70–107.

– and J.R. Vergnaud (1980) "Specifying Reference to the Subject: French Causatives and Conditions on Representations," *Linguistic Inquiry* 11: 97–202.

Safir, K. (1981) "Inflection-Government and Inversion," *The Linguistic Review* 1: 417–67.

– (1985) *Syntactic Chains*. Cambridge: Cambridge University Press.

Schmerling, S.F. (1973) "Subjectless Sentences and the Notion of Surface Structure," *Papers from the Ninth Regional Meeting of the Chicago Lingistic Society*.

Schmidt-Radefeldt, ed. (1976) *Readings in Portuguese Linguistics*. New York: North Holland.

Sportiche, D. (1981) "Bounding Nodes in French," *The Linguistic Review* 1: 219–46.

– (1983) *Structural Invariance and Symmetry in Syntax*. Ph.D. diss., Massachusetts Institute of Technology, Cambridge.

Steriade, D. (1980) "Clitic Doubling in the Romanian *wh*-Constructions and the Analysis of Topicalization," *Papers from the Sixteenth Regional Meeting of the Chicago Linguistic Society*.

Stowell, T. (1981) *Origins of Phrase Structure*. Ph.D. diss., Massachusetts Institute of Technology, Cambridge.

Strozer, J. (1976) *Clitics in Spanish*. Ph.D. diss., University of California, Los Angeles.

Taraldsen, K.T. (1978) "On the NIC, Vacuous Application and *that-t* Filter." Distributed by the Indiana University Linguistics Club.

Tellier, C. (1986) "Des quantificateurs flottants en français standard, en québécois et en mooré," ms. McGill University, Montréal.

Torrego, E. (1984) "On Inversion in Spanish and Some of its Effects," *Linguistic Inquiry* 15: 103–29.

Travis, L.D. (1984) *Parameters and Effects of Word Order Variations*. Ph.D.diss., Massachusetts Institute of Technology, Cambridge.

Tuller, L. (1986) *Bijective Relations in Universal Grammar and the Syntax of Hausa*. Ph.D. diss., University of California, Los Angeles.

Villiard, P. (1985) "INFL et nominatif," *Revue de l'association québécoise de linguistique* 4, no. 4: 169–78.

Vinet, M.-T. (1979) "Dialect Variation and a Restrictive Theory of Grammar: a Study of Intransitive Preposition in a Variety of French," *Recherches linguistiques à Montréal* 13: 107–25.

Williams, E. (1978) "Across-the-Board Rule Application," *Linguistic Inquiry* 9: 31–43.

– (1980) "Predication," *Linguistic Inquiry* 11: 203–38.

– (1981) "Argument Structure and Morphology," *The Linguistic Review* 1: 81–114.

– (1986) "The Italian Null Objects," ms. University of Massachusetts, Amherst.

Zagona, K. (1982) *Government and Proper Government of Verbal Projections*. Ph.D. diss., University of Washington, Seattle.

Zribi-Hertz, A. (1984) "Orphan Prepositions in French and the Concept of Null Pronoun." Distributed by the Indiana University Linguistics Club.

Zubizarreta, M.-L. (1981) "Subject Extraction in Portuguese and the Pro-Drop Parameter," ms. Massachusetts Institute of Technology, Cambridge.
– (1982a) "Theoretical Implications of Subject Extraction in Portuguese", ms. Massachusetts Institute of Technology, Cambridge.
– (1982b) *On the Relationship of the Lexicon to Syntax*. Ph.D. diss., Massachusetts Institute of Technology, Cambridge.
Zwicky, A.M. and G.K. Pullum (1983) "Cliticization vs. Inflection: English *n't*," *Language* 59: 502–13.

Index